ABOUT THE AUTHOR

Michael Sexton is a journalist with the ABC. He grew up in Adelaide, Melbourne, Canada and Thailand before studying journalism at the University of South Australia. After beginning his career in Bendigo he worked in Adelaide and London, including three years with the BBC World Service. His television work has included sport, news, lifestyle, travelogues, documentaries and current affairs. His written work has appeared in *New Statesman*, *The Age*, *Inside Sport* and *Australian Cricketer*. He is currently the South Australian correspondent for the ABC's national current affairs program *The 7:30 Report*.

Silent Flood
AUSTRALIA'S SALINITY CRISIS

Silent Flood
AUSTRALIA'S SALINITY CRISIS

Michael Sexton

ABC
Books

Published by ABC Books for the
AUSTRALIAN BROADCASTING CORPORATION
GPO Box 9994 Sydney NSW 2001

First published September 2003

National Library of Australia
Cataloguing-in-Publication data
Sexton, Michael.
  Silent flood: Australia's salinity crisis.

  ISBN 0 7333 1275 6.

  1. Salinity – Australian. 2. Salinity – Environmental aspects – Australia.
  3. Soil salinization – Control – Australia. 4. Soils, Salts in – Australia.
  I. Australian Broadcasting Corporation. II. Title

631.4160994

Cover photograph courtesy Jean Paul Ferrero/Auscape
Maps and illustrations by Ian Faulkner
Designed by Avril Makula, GRAVITY AAD
Set in 10.5pt Berkeley Book
Prepress by Colorwize, Adelaide
Printed in Australia by Griffin Press, Adelaide

5 4 3 2 1

For Jack and Cate and your generation

CONTENTS

ACKNOWLEDGMENTS

A work such as this could not have been considered without the help and encouragement of a number of people. I am indebted to many and in various ways. First to scientists and academics who patiently explained things time and again—and then took extra phone calls and emails to explain it one more time. To landholders, farmers and local government people who took me to salty areas and endured all the questions they had heard before and answered them willingly. To the bosses of large government departments and organisations who took time to explain the big picture and to Eric Rolls for a delightful and inspiring afternoon.

This book began as a documentary series for ABC-TV Education in conjunction with the National Dryland Salinity Program and its partners. So my colleagues also have my thanks and gratitude. Barry Mitchell who got it off the ground, Robert Clark who got me involved and Jonathan Holmes and Kerry O'Brien who granted me a sabbatical. The talented combination of Greg Ashman and Dean Heneker were the eyes and ears respectively of the programs. We travelled thousands of kilometres from Perth to Sydney to Launceston and a lot in between. The journey was in good company. Bob Lawrence had the task of looking at the resulting 50 hours of video and deciding how to reduce it down to two and his adjudication was terrific. For that and the steak sandwich, eight ball, pale ale, box trifecta lunches, I thank him. Others deserving credit include David Wenham, Kym Jannusch, Dianne Gall, Chris Moon, Bernard Humphreys, Ian Henschke and Martin Davies.

My thanks to those at ABC Books: Stuart Neal, Jacqueline Kent and especially Jenny Mills and Cheryl Rose for their patience and skill. I would also like to thank David Bevan for his kind words and fine example.

My gratitude to Liz cannot be contained in words and in writing this book my debt to her has grown even larger.

This book was conceived in education and so is not designed as a polemic. I believe every Australian in some way contributes to salinity and the better we understand the problem the sooner we can get on top of it. For those reasons I have not interviewed any politicians for this work—also such a work already exists. For those wanting a detailed look at the politics of salinity and water use in Australia I commend Ticky Fullerton's *Watershed* published by ABC Books in 2001.

A website built to accompany the TV series contains lesson plans and other information that may be helpful to teachers. See www.abc.net.au/learn/silentflood/

A number of articles from *SALT* Magazine have been reproduced in this book with the kind permission of the National Dryland Salinity Program and Currie Communciations. These are on pages 65, 103, 108, 128, 148, 158 and 167.

Thanks also to the New South Wales Department of Land and Water Conservation for permission to reproduce their glossary. The information on salt-tolerant trees and shrubs was reproduced with permission of the Queensland Department of National Resources and Mines.

Thanks also to ETT Imprint, Sydney for their permission to reproduce 'Scribbly Gum from Judith Wright's 'The Two Fires' in *A Human Pattern: Selected Poems*, and John Williamson and Emusic Pty Ltd for the use of his song lyrics, 'Mallee Boy'. The Pearl Buck quote is reproduced with the kind permission of Pearl S. Buck International.

INTRODUCTION

When I set out to document salinity in Australia I was told to head west. I was told that this was the place where people were most aware of the problem and where management solutions were most advanced. From the breathtaking destruction salt has caused to the most advanced techniques for halting its spread, the wealth of knowledge from that state is reflected in this book.

Of course while salinity is most obvious in the west, salt is also making destructive in-roads into the Murray–Darling Basin and towns and cities across the nation. In our egalitarian way we have all contributed to the cause of salinity and we will all pay some price—whether as a farmer facing crop damage or as a suburban family paying more for their food.

All levels of government across the country (especially the federal government via its $700 million National Action Plan for Salinity and Water Quality) are spending money fighting salinity in ways such as local solutions to urban salinity, subsidies for sustainable farming, research and development and monitoring the performance of entire catchments. The problem is so widespread though, that government funds alone could never cover the cost of salinity prevention and land restoration. In 2000 the National Farmers Federation and Australian Conservation Foundation estimated it would cost $65 billion over ten years to repair the damage already caused to the nation's landscape. Enormous parts of Australia will never recover and land, water and property will continue to be ruined—but there are also victories being won.

In this book I talk to the people who are dealing with salinity on a day-to-day basis and who understand both the problem and its management options. I've included expert opinions from a variety of perspectives in the 'viewpoint' boxes and a variety of success stories in the 'case study' sections.

While in Perth I met Shirley de la Hunty. To another generation she was Olympic golden girl Shirley Strickland who won medals at the games in London, Helsinki and Melbourne. Today she is a respected environmentalist, scientist and local government representative. Having grown up on a farm that faced salinity, she has one of the most all-encompassing views of the problem in the country. She talked about salinity with a steely urgency: 'Having known both the difficulties and joys of farming, I am absolutely certain that unless the whole community gets involved we will not beat it. It has to have full community support. Some people assume it is quite impossible to stop. I will not take that view because we dare not take that view. We must defend it. We cannot sit around and allow what might be a third of the country to become unproductive. It has to be handled and it has to be handled in a bipartisan situation.'

It is in that spirit that I have written this book. It would be impossible to answer every question about salinity in Australia, so I have attempted to logically tell the story of salt from its ancient origins to its current harmful effects. Salinity is like a silent flood because the build-up of salt mostly can't be seen or even imagined and the forces that drive it out are slowly and quietly building all the time. Once it reaches a critical level and takes hold on a large scale, it is almost impossible to hold back.

What lies ahead requires forward-thinking people on the ground and the best that engineering and environmental science can offer. It is a battle where the rules of engagement are different in almost every paddock and every catchment, the enemy is difficult and complex and it is all carried out with the disturbing knowledge that every day Australia's water and land is getting saltier.

Michael Sexton
Adelaide
September 2003

CHAPTER 1
In the Beginning

This closed all my dreams ... with bitter feelings of disappointment I turned from the dreary and cheerless scene around me.

EDWARD JOHN EYRE, 1840

When Edward John Eyre set off in 1840 to explore the centre of the Australian-continent he dreamt of finding an inland sea and country suitable for grazing. Instead he discovered a parched, desert landscape that was seemingly lifeless. As he stared ahead in dismay at the shimmering salt lake that would later bear his name, Eyre could not have imagined that it covered almost 10 000 square kilometres and contained one of the most remarkable ecosystems on earth.

In his travels north from Adelaide, the 25-year-old explorer had already come across another seemingly endless salt lake which he named Lake Torrens and described as a 'broad glittering strip'. His dreams already somewhat shattered, Eyre continued northward for another couple of hundred kilometres only to strike the southern (and smallest) section of Lake Eyre, which he believed was an extension of Lake Torrens.

Eyre had seen enough—he went home leaving behind a vestige of his feelings in the symbolically named Mt Hopeless, a hill from which he viewed the futility of his ambitions.

Over the next 40 years a series of explorers weaved through the inland areas Eyre had discovered and while each pushed on further, few had good things to say about Lake Eyre. Peter Warburton, after whom Warburton River to the north is named, called the northern section of the lake 'terrible in its death-like stillness and the vast expanse of its unbroken sterility', while Charles Sturt diplomatically wrote that it was 'a country which I firmly believe has no parallel on earth's surface'.

Sturt may well have been right. Lake Eyre is one of the great Australian misnomers because it is normally dry as a bone. The lake is known to have been full only about three or four times in the last century but when it is, it's a site almost beyond description. Its catchment area is roughly the size of Queensland taking in one-sixth of the Australian continent. It is inevitable water will find its way into the lake because at 15 metres below sea level it is the lowest point in Australia—the continent's outback sump. After rain in the tropical north, the outback rivers and creeks in Queensland's Channel Country swell into torrents of white water. The Cooper Creek explodes with fast-flowing waters as deep as 10 metres and when its banks burst the flow can be 50 kilometres wide. The Diamantina and Georgina and the Warburton rivers channel the wild water through the northeastern deserts of South Australia. Finally the water tumbles into the north of the lake and spreads thinly across its expanse forming a peaceful mirror of brilliant aqua. When the level rises high enough, a small neck called the Goyder Channel opens up and the southern lake begins to fill. When it filled to capacity in 1974, the highest recorded level, it was estimated to contain 34 cubic kilometres of water.

Within hours of the water arriving at the dried-out lake bed, the aquatic life emerges. Eggs that have been buried in the lake crust begin hatching and fish washed down in the torrents start breeding. Then the birds arrive. From flighty outback zebra finches to pelicans that have soared in from the Coorong hundreds of kilometres to the south, it's estimated one hundred species of birds find their way to the lake. How do they know there's water there? No one has the definitive answer but the theory is they sense the change in atmospheric pressure.

As soon as the rivers and creeks die down again and the water stops flowing into the lake system, the angry, outback sun starts the evaporation process. The annual rainfall for Lake Eyre is less than 125 millimetres but the annual evaporation rate is 2.5 metres—200 times greater. Within months the water level starts to drop and the water itself becomes brackish and then salty. Those birds that haven't already left or are too young to fly start to starve as the food chain loses its links. Eventually the muddy ponds give way to a hard, white crust. It has small ridges rippling across the surface where the wind has caught the last of the water and its final waves are frozen in time—or actually in salt. After the rain has come and gone there is just an immense sky, a bleak desert and 400 million tonnes of salt.

Welcome to the dead heart.

THE DEAD HEART

The great salt lakes of central Australia tell an important story both about the country and salinity. They show dramatically this most basic lesson—that if you visit the lowest land areas on the Australian mainland, you'll find salt. As the continent was evolving, geological processes caused an uplifting on the eastern edge and the centre started to sink. As a result, an internal drainage system was created that is unlike any other on the earth (*see box, Mary White p4–5*).

Australia's salt lakes have not always existed; there was a time when the centre of Australia was anything but a dead heart. Edward John Eyre was on the right track when he went looking for an inland sea—it's just that he was 110 million years too late.

When Tim Flannery, the Director of the South Australian Museum, is asked to describe what central Australia looked like during that evolutionary period, his eyes light up and a childlike smile of enthusiasm creeps across his face. 'These stones that you see in Australia's centre were brought into that area by glaciers

VIEWPOINT
Mary White

MARY WHITE IS A PALAEOBOTANIST AND THE AUTHOR OF
FOUR BOOKS THAT TOGETHER DOCUMENT THE EVOLUTIONARY
HISTORY OF AUSTRALIA'S LANDSCAPE AND ITS MORE RECENT
DEGRADATION. HER WORK HAS EARNED HER A DOCTORATE OF
SCIENCE FROM MACQUARIE UNIVERSITY AND THE QUEENSLAND
UNIVERSITY OF TECHNOLOGY.

*Australia is different from any other land that I
know of. When you think about a river, it usually
builds on high ground and runs briskly down slopes and
disappears into the sea. When rivers behave like that, as they
do around the coastal parts of Australia, any excess sediment
from erosion and excess salts that actually enter the river
systems are carried back into the sea.*

*That's all very fine but Australia mostly doesn't behave like
that at all.*

About 18 million years ago you had the tilting of the edge of the continent, which formed the Great Dividing Range. Most of our major rivers rise on the western flanks of the Great Divide. So they don't run briskly down slopes and into the sea, they run into a very flat landscape.

In the case of the Murray–Darling it's our major river system and after travelling mainly westward, the two rivers join and then have one very small exit to the sea. So you've got inward-flowing rivers in a continent with a sunken centre. This is an amazingly different drainage pattern where all the central drainage on the continent flows towards Lake Eyre.

When rivers don't carry the sediment off the land, and it is retained, the landscape gets flatter and flatter and becomes a floodplain. At the same time of course, it also retains its salt.

As Europeans arriving in Australia we thought it wouldn't take long to change the environment to the sort of landscapes we were used to and we imposed on it the sort of agriculture and land use practices that suited the northern hemisphere.

But the environmental history of this land has been so utterly different from the northern hemisphere. Add to that the soils are old and originally made from weathered rocks and you have some concept of the difference. Of course when we were colonising this place and starting all our farming activities nobody had any appreciation of the amount of salt that was held just below the growing region of the soil, which was just waiting to cause problems when we altered the hydrology.

110 million years ago. Back then there would have been an inland sea with ice, pack ice and icebergs brought by glaciers off highlands. There would have been forests growing around the edge of the inland sea but probably forests like you would find in Alaska today, quite limited in numbers of species. In that frigid water would have lived enormous marine reptiles. They weren't dinosaurs although they were from the age of dinosaurs. They were ichthyosaurs that were large animals that looked like dolphins, except maybe they were 7 or 8 metres long, and quite deep-diving animals.'

There are those who believe that this ancient inland sea is responsible for leaving a load of salt that is causing problems today but they are in the scientific minority. What remains are fossilised shells, molluscs and bones. The salt is a recent arrival compared to these subterranean treasures.

Scientific studies have confirmed the existence of an ancient interior landscape that was abundant in plant life and inhabited not only by reptiles but also by mammals. In 1893 the South Australian Museum sent a palaeontologic expedition to explore the far north of the colony. Like most travellers to this arid part of the world they were initially speechless at what they saw. When they looked out across the rock-hard saltpan of Lake Callabonna on the edge of the Strzelecki Desert they saw hundreds and hundreds of bones bleached white in the sun. These were not the bones of kangaroos or camels, but the remains of herds of diprotodons—a hippopotamus-sized marsupial that grazed for food presumably in verdant environments. The diprotodon was part of what is known as megafauna, of which about 60 species, including ancestors of kangaroos, emus and wombats, are thought to have existed.

Almost a century after the discovery at Lake Callabonna a Sikh scientist, Gurdip Singh, confirmed the area had indeed been lush. Dr Singh had pioneered a technique for reconstructing pieces of pollen that were preserved in moisture under salt lakes in Rajasthan in India. In the 1970s and 1980s he visited the salt lakes of central Australia and found pollen samples under the crust. Through reconstructions he was able to unlock the history of the vegetation of inland Australia. In short, he showed that far from being a dead heart, this region had been a changing environment that had at times, supported lush vegetation.

So how did it get to be so salty?

THE ICE AGE DID NOTHING FOR US

The diprotodon whose skeletal remains were so beautifully preserved at Lake Callabonna seems to have died out only about 40 000 years ago during a major ice age (Pleistocene). This ice age, which began about 3 million years ago, was the last great period of upheaval for the globe and its human population, but one that did Australia few favours and ultimately contributed to the salinisation of the country.

While the earth was again going through one of its cool phases, enormous ice sheets covered most of Europe, North America and Asia. Although it was cold it was also dry and windy. The fresh water froze and the polar caps expanded. The sea levels dropped and there was great human migration as groups walked across temporary land bridges. This is the period when Aboriginal Australians moved from Asia onto the great southern continent. When the ice age ended, about 10 000 years ago, the great melt began. Sea levels started to rise and rain began falling again.

For the countries most affected by the ice age, a beneficial side effect was a gift of new topsoil. This was created by the grinding of rocks under the enormous pressure of heavy, slow-moving ice sheets. There was also much seismic activity. Volcanoes spewed lava and earthquakes churned up the country leaving behind fresh, fertile land.

The shame is that virtually none of this happened in Australia. Apart from some glaciers and a few notable volcanoes (the extinct Undara volcano in far north Queensland and Mt Gambier volcano in South Australia) there was little geological disruption. Australia's climate became drier and windier and its topsoil became older and wearier.

The nutrients that were being tossed around abundantly in Asia, Europe and North America were dormant here. The winds that hammered away blew sand, soil and salt in all directions. Although

WHAT IS SALT?

Salt is probably the most recognisable mineral on earth for one very good reason: it tastes salty. For the purposes of chemistry what we call 'common salt' is 60.663 per cent chlorine (Cl) and 39.337 per cent sodium (Na) that combines to form sodium chloride (NaCl).

Salt is easily created, which is a good thing given how important it is to life. The human body cannot create salt but it depends upon a certain amount of it for survival. Salt has been given a bad name because of medical research showing too much can put a strain on the blood circulation system. This hypertension is responsible for heart and kidney disease and strokes.

However, without a certain amount of sodium in the system, the human body cannot function properly. The sodium is an electrolyte, meaning it carries a small electrical charge and this is needed for nerves and muscles to operate properly. Chloride is used by the body to help in digestion. The average adult is thought to need a minimum of 0.23 grams of sodium in their system to function properly. This is equivalent to 0.57 grams, or less than one-tenth of a level teaspoon of salt.

Humans aren't the only creatures that crave salt. Animals, particularly herbivores, will seek out salt sources. That is the reason why salt occurs in red meat. Farmers for generations have been known to give domesticated livestock a salt block to lick to satisfy the animal's desires.

Natural salt occurs either in crystals, which are rather like tiny cubes, or dissolved in water. The purest salt (99 per cent NaCl) is made from evaporated seawater (sea salt). There is also halite or rock salt that is found in sedimentary layers of the earth. This is usually about 95 per cent NaCl and has been formed by seawater being trapped underground.

However in terms of salinity in Australia, salt is mostly just common salt.

Australia's earth was being redistributed it wasn't being renewed: at the end of the ice age Australian soil had simply aged another couple of million years. (In comparison some of the soils of Europe are only 25 000 years old.) In central Australia the landscape had become mostly desert; elsewhere it was thin and fragile and vulnerable to among other things, salinity.

MOON PLAIN

In 2001 a group of scientists from the South Australian Museum made a trek to the far north of that state. A potter from Coober Pedy and his son had stumbled across some ichthyosaur bones while they were out riding their motorbikes in an area known locally as Moon Plain. There is a good reason for the lunar comparison because the landscape looks almost nothing like any other part of planet earth. Tim Flannery laughs when he remembers the first time he visited the site to check on the remains. 'Just when you think things can't get any bleaker, there is Moon Plain.' Dr Flannery calls the landscape 'sterile' and aside from the fossilised shells, wood and bones from the ancient inland sea, there is nothing else. It is so barren and quiet you can almost hear the heat baking the ground.

Each day for several weeks the museum staff drove out to Moon Plain from Coober Pedy and spent the day on their hands and knees. They used paintbrushes to dust off the precious fossils before noting them and packing them carefully into boxes to take back to Adelaide. However one day on their way out they diverted from their usual route and unexpectedly came across the most dramatic illustration of soil degradation in Australia. Amid the flat landscape was a mesa that was almost black but for the dazzling bright reflections it emitted. As they approached the mound they felt they were looking at an ancient dumping site where the only rubbish was thousands of panes of glass. These panes were up to 2 centimetres thick and some were a metre long. They jutted out at all angles and seemed to be buried randomly throughout the area. Climbing up the mesa's slopes was like trying to walk through a fresh snowfall. The scientists' boots sunk into the ground as it gave way beneath them, making a crunching sound like stepping on dried leaves. Palaeontologist Ben Kear says the panes were not glass but gypsum. 'Gypsum is an evaporative mineral which has come through the sediment and formed within gaps. It forms these big plates of glass, I suppose you'd call it, within the sediment and that weathers out to the surface. It would suggest there was a lot of salt around but it's a result of extreme aridity drawing all the water out of the sediment and leaving this dry barren landscape we see today.'

The group found no life forms in the area—not even the most basic plants. The soil is so old and weathered that it almost had no structure left. Moon Plain is almost 500 kilometres from the coast but it is an area that contains tonnes of salt. This may not be the 'endgame' of Australian soils but it is a good place to start looking for it.

A BREADBOARD FLOATING IN A SALINE BATH

'That's what we are like,' says John Williams from CSIRO Land and Water Division while holding out his hands, flat palms facing up, 'we are like a breadboard floating in a saline bath. Our land is old, hard, weathered and dry and we are surrounded by ocean.'

It is the simplest explanation to the question: Why is there so much salt in Australia?

As anybody living near the beach will tell you, salt comes in from the coast. All maps of the world are dominated by the blue oceans and seas and their power remains a driving force on the planet. Covering 70 per cent of the earth's surface they contain 80 per cent of all life forms. They create our weather and are the source of all water.

And dissolved in that water is salt. The Smithsonian Institute in America estimates there is 50 million billion tons (45 million billion tonnes) of it. Enough so that if it were collected and heaped on the land, every continent would be covered to a height of one-and-a-half metres.

No country is more affected by this salt than Australia because as an island continent it's surrounded by three oceans and four seas. So how does salt from the ocean enter our land? The answer is in the weather. This country gets saltier with every breath of wind and every drop of rain.

As seawater evaporates it brings with it tiny amounts of salt that later fall to earth in raindrops. While a raindrop may carry a tiny grain of salt hundreds of kilometres inland, once it falls to earth it mostly stays there. Unless it collects in a creek or river and heads quickly to an ocean outflow, the rainfall is usually absorbed deep down into the soil and the salt will go with it. Any evaporation from the ground will take the moisture but a vast majority of the salt will stay behind in the soil. The salt is in for the long haul.

Although rainfall is one way salt gets into the Australian landscape, most of it comes from another weather-borne source—the wind. Dr David Chittleborough from the University of Adelaide says scientists mostly agree that a majority of Australia's salt has been blown in from the oceans. 'Because Australia wasn't swept clean by the ice sheets 10 000 years ago as it was in the northern hemisphere, we've had soils that have been accumulating salts for hundreds of thousands of years. If you do your sums you can soon account for a lot of the salt that's in the Australian landscape simply from wind-blown accession from the oceans. In coastal regions it can be as high as 200 kilograms per hectare per year while inland it is reduced to about 20 kilograms annually.'

John Williams explains what has happened once the salt arrives on an accumulating landscape: 'There have been periods of time when the climate has been very, very dry. During those dry times salt that has accumulated on the surface and in the salt lakes was picked up and blown back over the landscape, often in places where that salt had long since been moved out. So we've got a landscape that is flat and salty, and the distribution of the salt in that landscape is quite complex.'

It is also the reason why we can't get rid of the stuff.

SLOWLY DRAINS THE SALT

A popular saying used whenever environmental matters are discussed is that Australia is the driest continent on the planet. Antarctica

THE WATER CYCLE

The oceans and seas contain 97 per cent of the world's water but contained in that water is an estimated 45 million billion tonnes of salt. Ocean salinity is the result of millions of years of a process known as the water, or hydrologic, cycle. This circular movement of water from oceans onto land via rainfall and back via runoff is one of the driving forces of life on planet earth.

The sun evaporates seawater where it forms into clouds. Because of the high salinity level in the ocean, as the water is drawn up, it brings some salt with it that remains in the cloud. When the clouds run into a stream of cooler air—usually over land—precipitation results and the condensed moisture becomes rain, sleet or even snow and falls to earth. Tiny particles of salt fall to earth at the same time.

Once it has landed, the water can go several ways. It can run back into the sea or ocean, it can run into rivers or creeks, or it can soak into the earth and become groundwater where it seeps down into the watertable (the upper level of saturated ground). The cyclical process continues as the sun continues to pull water back up into the atmosphere. Where water is exposed, such as in puddles, creeks, rivers or dams, evaporation occurs. Water that soaks down into the ground isn't immune though. The power of the sun can evaporate moisture from as deep as 2 metres below the surface. Water also returns to the atmosphere from plants in what is known as transpiration. Tiny openings in the leaves of plants allow moisture that has been drawn up from the ground through the root system to evaporate. Water returning to the atmosphere in these ways is known as evapotranspiration and although the water is taken up it leaves behind most of the salt it brought in from the sea.

The process of water percolating down underground into the watertable is known as recharge (or recharging the watertable). Discharge from the watertable is when the volume of water

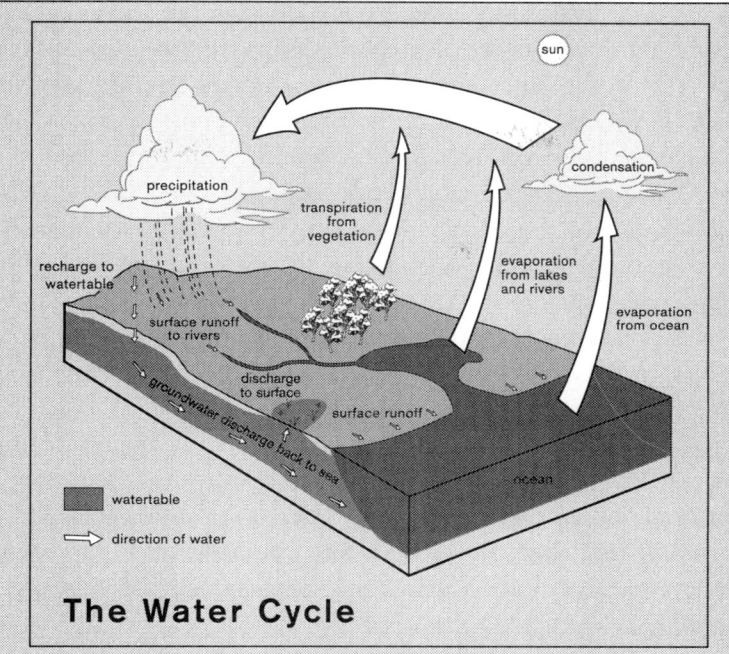

The Water Cycle

builds underground to the point where it reaches the surface and is forced out into low-lying land, creeks, rivers and oceans.

Australia's unusual geography creates a slightly distorted hydrologic cycle. With the continent being so flat and draining internally, the water takes much longer to reach the coast—if at all. There are few fast-flowing rivers that deposit salt quickly back into oceans or seas. The evaporation taking place will take only some salt with it; most of it is retained in the landscape.

Oceans vary in salinity levels depending on where they are on the globe. Three oceans and four seas surround Australia and they vary in salinity levels. There is relatively less salt in the Coral, Arafura and Timor seas to the tropical north because of the higher rainfall and proximity to land. In the Tasman Sea and Southern, Pacific and Indian oceans levels of salt are higher up until a third of the way towards Antarctica where the increased cloud, rain and proximity to the polar cap dilutes the salt levels.

actually takes that prize. However it is not aridity, but something less superficial that goes to the heart of our salinity problem. Australia has 5 per cent of the earth's landmass but only 1 per cent of total river runoff. Given that one-sixth of the continent drains internally into Lake Eyre and the high points of the country are generally near the coast, Australia has a serious drainage problem.

We know salt will stay dissolved in water unless there is evaporation. If the water drains away it takes the salt with it. In well-drained areas of the world such as New Zealand, salinity never sets in because of the speed with which groundwater finds its way to the coast. Unlike Australia, where the greatest river system, the Murray–Darling, mostly doesn't flow so much as amble out to sea. More water flows out of the mouth of the Amazon River in a day than from the Murray in a year. Since European settlement this lack of water is mostly the result of extraction upstream but historically the Australian continent doesn't give up water easily. Only 10 per cent of Australia's rainfall ends up as river discharge.

The problem with having little drainage is that the saline water sits in the ground waiting to seep out to sea, and while it sits there, it is possible for the salt to come to the surface and cause salinisation. This process occurs regularly in parts of the country where evaporation rates are high. This salinisation has been going on slowly for thousands of years and is known as primary salinity. It shows itself as salt marshes, salt flats and salt lakes. Australia isn't the only place with it. All over the world where similar conditions occur there is primary salinity. The Western Australian Department of Agriculture estimates Australia has about 29 million hectares of primary salinity. It divides that figure into two categories: the first is marshes, flats and lakes (14 million hectares) and the second is naturally saline subsoils (15 million hectares) where no groundwater or perched watertable is able to take it to the surface. There is no difference in the process of primary salinity and dryland or irrigation salinity—it's just the conditions that dictate how fast it develops. Farming methods have dramatically accelerated the salinisation of the continent but it is wrong to think salinity began with western agriculture. It was around long before the first tree was felled and there are plenty of Aboriginal and colonial anecdotes to prove it.

SALT STORIES

Given that the heaviest falls of salt are near coastal areas, it is not surprising that Tasmania once had large natural salt deposits and soaks. Kay McPherson, an Aboriginal elder with wild hair and sparkling eyes, has tracked down the history of this salt. She has a fascination with geology and how it fits into the history of her state and its people. Her studies show there were significant collections of salt in Tasmania that resulted from primary salinity (*see box, Kaye McPherson p18–19*). This spawned an industry on the island which Kaye believes led to early racial skirmishes in Van Diemen's Land.

Colonial history on the mainland suggests the Murray–Darling Basin also had its share of primary salinity. In 1828 New South Wales governor Charles Darling sent Captain Charles Sturt and Hamilton Hume into the western interior of the colony to trace the Macquarie River and to see if there was an inland sea. Their timing wasn't brilliant because the region was suffering from a severe drought that had reduced creeks to dusty channels. Sturt records how the party was struggling in the heat and was almost delirious when they came across what he described in his diary as 'a noble river' which was 80 yards (73 metres) across and covered with pelicans and other birds. He would name the river after Governor Darling and record in his diary a vivid description of early salinity in Australia ...

> Its banks were too precipitous to allow of our watering the cattle, but the men eagerly descended to quench their thirst, which a powerful sun had contributed to increase; nor shall I ever forget the cry of amazement that followed their doing so, or the looks of terror and disappointment with which they called out to inform me that the water was so salty as to be unfit to drink! This was, indeed, too true: on tasting it, I found it extremely nauseous, and

strongly impregnated with salt, being apparently a mixture of sea and fresh water. Whence this arose, whether from some local causes, or from a communication with an inland sea, I knew not, but the discovery was certainly a blow for which I was not prepared. Our hopes were annihilated at the moment of our apparent realization. The cup of joy was dashed out of our hands before we had time to raise it to our lips.

Clearly Sturt, Hume and the party had stumbled across a brackish Darling River that had insufficient water to dilute natural salt springs and soaks. Future generations would tame the river with engineering and irrigation. This would often be cursed as the creator of salinity and there is truth in that—but it is worth remembering there has always been a lot of salt in the system.

Prior to colonisation the rivers of southern Australia were in a boom–bust cycle that saw them fluctuate from the broad surges of wild water to a string of billabongs. River trade on paddle-steamers and later irrigation farming led to demands for more consistent flows of the rivers. To smooth out the peaks and troughs of water levels, an elaborate system of locks, weirs and dams was built along the major waterways. Between the 1920s and 1940s that would change the water flow forever (and keep the salt and fresh water separate at the river's end where it runs into the southern ocean at Goolwa in South Australia). This gives the impression that there is a natural and predictable body of river. However, a quick wander through the archives or libraries of country towns in the Murray–Darling Basin will usually show extraordinary black-and-white photos of the impact of floods and droughts in the region. While one shows people holding a picnic in the dry riverbed in front of the Renmark Hotel, another has a man rowing a boat past the top of his farmhouse chimney.

The legendary Murray cod, a fish that can grow well over a metre long and lives up to 75 years, is testament to a species' adaptation to the historical boom–bust cycle. The cod builds enormous stores of fat that can sustain it for up to seven years without eating. So if it is caught in a drought cycle, it can sit it out as the river is reduced to a series of puddles. When the rain comes and the river flows it begins eating again. Having evolved from ocean fish the Murray cod can also tolerate salt levels rising in the water. Just as well because when the droughts came the system salted up and when the water flowed again the system was flushed. In big floods it is estimated the Murray mouth was 10 kilometres wide and the fresh water would leave a muddy mark on the ocean floor for 20 kilometres out to sea. Although the continent drained slowly, the boom–bust cycle ensured that the salt could make its way out.

Western Australia has generally been known as the nation's home of salinity and according to the earliest historic records it seems the title is well deserved. The colonies on the eastern seaboard were already 40 years old when the Swan River colony was established. The British settlement was mostly a tactical move to thwart the French who had been considering the west for some years. They weren't the only European powers interested in the region—the Dutch had also navigated the coast. Historical evidence suggests that salinity played a part in delaying settlement, particularly with the conservative French. It seems the French struggled so often to find fresh water after landing in Western Australia that some hypothesised the local indigenous people had evolved to be able to drink seawater. This bizarre explanation came after creeks and soaks all proved too salty to drink and the visitors were forced to distil seawater to survive. The Swan River colony battled constantly for fresh water, particularly in summer, and inland migration was slowed by the curse of salt. From 1839 when the first explorers started to walk what would later become the wheatbelt, they described pools of brackish water and large areas of salt lakes. They also noted the vagaries of water and rainfall in the region. The colonials would have had no idea that the southerlies roaring across Western Australia had been dumping tonnes of salt onto the land for hundreds of thousands of years.

The evidence is overwhelming that primary salinity was embedded in Australia before European settlement. When the colonials started arriving they struggled to come to terms with the land and its inhabitants. Colonial art gallery collections are full of

VIEWPOINT
Kaye McPherson

KAYE MCPHERSON IS A TASMANIAN ABORIGINAL CULTURAL HISTORIAN AND GEOGRAPHER. SHE HAS A SCIENCE DEGREE SPECIALISING IN GEOGRAPHY AND ENVIRONMENTAL STUDIES. HER MASTER'S RESEARCH AT THE UNIVERSITY OF TASMANIA IS ON THE IMPACT OF EARLY INDUSTRIES IN TASMANIA.

>
> *There is a lot of documentation about Aboriginal use of salt but like all Aboriginal history in Tasmania it is coming out of the European account.*
>
> *The Aboriginal people wore fur coats and the furs were tanned. They had sinews they wore as jewellery so I would imagine they were using salt for all that. I know they didn't use salt so much as a condiment because they used white ash from the fire for that purpose.*
>
> *A lot of the saltpans here were massacre sites. When you look at a massacre site you need to ask why Aboriginal people were coming in there. They were coming in to use the salt that*

*was being controlled by European people. Therefore
Aboriginal people weren't allowed to use it.*

*Through the midlands of Tasmania you get a lot of salt in
areas associated with what is called the Ross formation [an
ancient sandstone formation]. In outcrops you will actually
get salt seeping out of the sheltered sandstone places where it
can be collected. I believe this was used as a source of salt
especially after the saltpans became degraded.*

*The first harvest of salt was recorded in the saltpan plains
[a saltpan 2 kilometres wide and 4 kilometres long in the
midlands east of Tunbridge] in 1809. Soon there was a regular
salt industry harvesting 100 tonnes per year. It had so many
uses including the tanning and preserving of meat. It was also
used to treat the wounds of convicts after they had been lashed.*

*The colonial government authorities discovered quite early
on that when the sheep went missing the convicts were actually
killing the animals and then getting the salt off the pans so
they could preserve and sell it in the towns. To stop that they
brought in licences and salt trading was actually run through
the local police who charged 10 shillings for a wagonload.*

*By the 1850s the salt industry had ended because the
saltpans were ruined. The wagons had been driven onto the
area to collect the salt and so had contaminated it. Some salty
areas also were lost because farmers began draining marshes
to improve the land and so the salt intake was lowered.
Distilling seawater either by natural evaporation
or boiling was then used to manufacture salt.*

paintings depicting gums that have been distorted into elms and oaks and kangaroos with dog-like features. Could they be blamed for trying to find something familiar in a place so unfamiliar? The Australian flora and fauna had evolved to cope with a continent that could be as cruel as it was beautiful.

When Edward John Eyre staggered down from Mt Hopeless and scrawled his dispirited note he could barely interpret what he was seeing. The horizon of salt burned his eyes. He didn't know the salt had been blown across the land by ferocious winds and rain and then swept up in the floodwaters of the northeast and deposited in central Australia. He couldn't have known the continent's prehistory had lifted the crusty edges of a hard-baked slab of country that sunk in the middle. In this heat how could Eyre imagine a place that was dry, cold and windy but without glaciers or volcanoes to turn and refresh the ground? What he saw was what he believed to be the gates of hell. If he could have turned his gaze away from the glittering, endless saltpan he might have noted the vegetation that did survive this heartless place. Because the way trees, in particular, grow and thrive in this degraded, saline soil and arid climate tells the most basic story about salinity in Australia.

CHAPTER 2
Let Her Rip

The gum-tree stands by the spring
I peeled its splitting bark
And found the written track
Of a life I could not read

JUDITH WRIGHT, 'SCRIBBLY GUM'

On both sides of the South Australian–Victorian border is a series of national parks. They aren't particularly fashionable and are unlikely to be photographed and splashed across glossy tourist brochures. These national parks are in mallee country where the soil is sandy and the bush is tangled. There are no creeks with mossy banks thriving beneath the giant canopy of rainforests. Here plants must be able to survive in country with low rainfall, burning sun and poor soil. The beauty of the mallee is modest but evident in the subtle colours of tiny flowers and in slow-moving animals such as wombats, shingle-back lizards and echidnas. The largest slab of mallee country covers the Ngarkat Conservation Park in South Australia and the Big Desert Wilderness in Victoria that abuts it. Together they cover more than 400 000 hectares. This is just a remnant of the belt of mallee that once covered half of Victoria, South Australia's southeast and Eyre Peninsula and the southwest of Western Australia.

Saving mallee country is not, historically, the sort of issue that makes people want to lie down in front of bulldozers. At times, even the word 'mallee' is used to describe a problematic environment: it is seen as difficult to clear and the trees useful only for burning. With poor soils to work with, mallee farmers face an uphill battle once they have cleared the land and started farming. They have a reputation for being as tough and enduring as the country they tame and Australians who have never even seen this part of their country still refer to others as being 'as tough as mallee bulls'. Singer–songwriter John Williamson grew up in the Victorian mallee and celebrates the culture in his song 'Mallee Boy'.

Where little town dogs howl at the morning train,
Where a cocky makes a living on twelve inches of rain
Where his woman provides and is rare to complain
And I still love the smell of that sandy soil,
Some say it's dusty, some say it's gold
Cause it grows the sweetest fat lambs the markets ever sold
And I don't mind at all if you call me a Mallee Boy,
No I don't mind at all if you call me a Mallee Boy.

The central feature of this country is the mallee tree itself—a plant that is the victim of some serious misinformation. Far from being troublesome, it is one of Australia's most astonishing pieces of nature. If the kangaroo and platypus are upheld as examples of how animals have developed unique adaptations to survive on this continent, then the mallee must be the flora equivalent. Its ability to survive in one of the harshest environments on earth is worth examining and, in doing so, discovering how this ancient continent survived salinity.

GROWING UP TOUGH

Palaeobotanist Mary White has spent much of her adult life researching the fossil records of Australian plants and trees. Similar to a social worker discussing a truant whose behaviour is the result of a troubled childhood, she explains the impact the prehistoric formation of Australia had on its soil and, in turn, the trees that grew up in it: 'The characteristic of our soils has always been ancient soils made from deeply weathered rocks over most of the continent. Therefore we are lacking in soil nutrients. Normally when you have forest or anything approaching heavy forest you have complete recycling of all the vegetable matter and a complete recycling of nutrients. So they are basically making their own better soil conditions. However the moment you get dryness coming in that doesn't apply anymore. You don't have the rotting of the leaf litter, you don't have the necessary microbiotic and invertebrate activity in the soil. Therefore you are stuck with the poor soils that characterise most of the continent. Our typical Australian vegetation is the result of the selection of the things that could cope with dryness and occasional fire.'

Nowhere is the soil poorer and the rainfall lower than in the country known as the mallee. A mallee tree can grow comfortably in areas that receive as little as 250 millimetres of rain per year. The mallee grows very slowly and densely, its foliage characterised by leathery leaves and occasional red flowers. It can survive long periods of drought and then burst into growth after rain because of the enormous larder it keeps below the surface. The mallee's lignotuber (stump) stores nutrients and grows buds that shoot when conditions are right. From the surface the tree will often look like a series of trees growing out of a single stump. The tree can grow for hundreds of years and when mature, its wood is so dense it will sink in water even when it is dry. That is why mallee logs and particularly the lignotuber is a favourite for home heating.

When asked about the mallee Mary White says: 'Like so many of our trees and different plants they are very finely tuned to the difficult conditions which are firstly poor soils and then a very uncertain climate. The mallee is a marvellous example of adaptation because the mallee is a eucalypt but it doesn't grow like a gum tree with a solid trunk forming like the sorts of trees you see in high woodlands. It grows a number of much lesser trunks much more like sticks in all directions and its main trunk structure is underground; it's that great mallee root system. This enables it to cope with drought, and if the top part dies off in a drought or is burned off in a fire, it starts growing from this great big woody structure. Like other eucalypts its leaves hang at an angle where they catch the least amount of sun and so minimise the amount of evaporation. Mallee is one of the things that is well adapted because it is in the marginal or semi-arid lands where ordinary gum trees simply wouldn't grow.'

The great mallee root system might be appreciated now but it was cursed when farmers were trying to turn the bush into fields and pastures. Beyond the lignotuber the root system curled deep into the sandy soil. Ripping the top trunks off a mallee tree was not good enough because the underground system would send more shoots up straightaway. The mallee stump had to be levered, or in some cases blasted out of the ground with dynamite to make sure it was completely removed—evidence of one seriously determined plant. But what connection does the mallee have to salinity?

EUCALYPT DRINKING STORIES

There is one very important fact about the mallee that helps us to understand its significance in the salinity story. A mallee tree is capable of absorbing 99.9 per cent of rainfall. Given that it grows in semi-arid conditions, this may not seem much of an effort. However it shows how this native tree has adapted to the climate and become extraordinarily efficient at using water and nutrients—of which it has very little. In areas where there are large salt deposits in the ground, the watertable turns saline as it absorbs the salts and becomes saline groundwater. A farmer who sinks a bore into such areas containing saline groundwater will bring brackish, or in extreme cases, saline water to the surface. A brackish or saline watertable is of course, a very unpleasant thing for a tree or other deep-rooted plant to experience. Few trees will tolerate having wet feet for long periods of time, let alone wet and salty feet. The best way for them to avoid a saline watertable is to use up the water before it turns salty. Say, perhaps, 99.9 per cent of the water.

The mallee tree by its efficiency saves itself from salty water.

The mallee forests of western Victoria grow along the state's border all the way to the Murray River and then continue north of the river into New South Wales. However before Europeans arrived in Australia there were also great forests of river red gums growing in the Murray–Darling Basin. Unlike the mallee's mutant form, this gum is a truly magnificent-looking tree that derives its name from the heavy, red timber encased in its grubby white-bark skin. The river red is not strictly a water plant but it can grow during occasional periods of flooding. A long-term study established in the mid-1980s by CSIRO, the Western Australian Department of Agriculture and the University of Western Australia has examined the drinking habits of various eucalypts at a test site near Katanning in the Western Australian wheatbelt. The research showed that the river red gum sends roots down to the saturated area just above the watertable to draw moisture and nutrients, as opposed to other gums that prefer to drain the drier soil near the surface. It would seem the river red gum knows just how far to go looking for moisture while avoiding salt.

River red gums are only one branch of about 500 members of the great eucalypt family that are notorious drinkers. When Israel was developing after World War II, the government in Tel Aviv bought thousands of the trees from Australia to help soak up marshes and other boggy land so it could be reclaimed for farming. In the late 1990s a similar program was under way along the Mekong River in Laos. It seems nothing is as thirsty as an Australian expatriate. Botanists know these eucalypts are capable of outdrinking other plants around them quite simply because they go deeper in search of moisture. The young trees start with a base of roots just below the soil surface and then send a taproot as deep at 15 metres down into the earth. In addition to reaching down vertically, the roots have been known to grow horizontally as far as 20 metres in search of a drink. Each day a tree loses hundreds of litres of water through

VIEWPOINT
Dr David Chittleborough

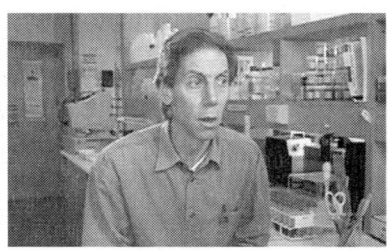

DR DAVID CHITTLEBOROUGH IS HEAD OF THE DEPARTMENT OF
SOIL AND WATER AT THE UNIVERSITY OF ADELAIDE. HIS RESEARCH
INTERESTS INCLUDE SALINITY AND SODACITY AND THE ORIGIN
AND DEVELOPMENT OF SALINE SOILS.

There are different types of salinity, but they are all having a fairly big impact on our agricultural systems and certainly the sustainability of our environment generally. The salt giving rise to so much of the problem that we see is largely contained within saline groundwater, which is within various depths of the soil profile. Sometimes it's near the surface, sometimes it's at some depth. Because we have stripped away a lot of the perennial vegetation and replaced that vegetation with shallow root crops, larger amounts of rainfall are eventually getting down into the ground through the soil profile and into the groundwater.

It is like filling up a glass that is half full. As you put the water in, it will gradually fill up and that's essentially what is happening at the moment and of course that is affecting the lower parts of the landscape.

What it does to the soil is quite damaging because it can saturate the soil with sodium. You end up with sodium on the soil surface and that causes the structure of the soil to be very degraded. Once you remove that salt through various means you are left with a poorly structured environment.

You can do an experiment in the laboratory where you can remove salt by adding various amendments like organic matter and calcium. This gets rid of the salt through leaching and you can produce quite an acceptable soil profile for plant growth. It depends a bit on the nature of the soil but you can do it. However when we are talking about hundreds of thousands of square kilometres including catchments of different sizes and shapes and vegetation types let alone the different land uses involved, you have a process which is way beyond what I have described in the laboratory. You just can't do that.

I think the Australian public should see salinity in its various forms as the most serious and intractable environmental problem that we have. It far outweighs any of the other environmental problems that this country has and it's going to require massive intervention in order for us to get on top of it, to live with it, to restore our landscape and have a sustainable future.

moisture evaporated from leaves and is replaced by sourcing water via its roots. A large river red gum can use over one thousand litres in a day.

The Western Australian study also looked at the yellow gum and coastal mort, which can both survive in soils too dry for most other plants. They found that although these trees couldn't access the deeper watertable, the leaf tissue altered during dry and hot times to allow growth and gas exchange to continue.

John Williams from CSIRO's Land and Water Division says the amount of water that is able to leak past the root system of native trees is commonly 1–5 millimetres per year. This thirst is wonderful protection against salinity because the best way to deal with salinity is to avoid it happening by using up the rainfall before it turns into salty groundwater. Although the classic images of saline country show a hard-baked saltpan in an arid setting, salinity is actually mostly about water. We might live on the driest vegetated continent on earth but Australia's rising watertable is a silent flood, and how we manage our water will decide how far the watertable rises.

Environmental historian Eric Rolls paints word pictures about the size and quality of landscapes that now are as tangible as dreams: 'The heath country in Western Australia was some of the heaviest in the world. It was so dense it never felt the wind. Yet it grew in some of the poorest soil in the world; it was a marvel of recycling. There were native grasses with roots up to 10 metres deep and they had as much or more effect on keeping the watertable down as the trees. There was a belt of grass country about 200 miles (322 kilometres) wide that ran from central Queensland all the way to Victoria. It was full of koalas. Less than 5 per cent of that is left in Australia.'

Before Europeans arrived very little rainwater reached the watertable. The system was balanced because the covering of trees such as mallee, river red gums, casuarinas and native grasses together soaked up most of the water and the watertable remained deep down in the earth. The salt that was absorbed in water didn't surface and was able to slowly leak out of the continent or into the sunken centre.

BIG LIZZIE

Julia Cotching is in her nineties but she still has a clear vision of the monster she saw as a seven-year-old girl in 1917. It was a creature part to be feared and part to be admired. When asked to tell the story she started by slapping her hands slowly up and down on the arms of her chair in a nursing home at Irymple in Victoria: *[slap slap slap]* 'All the kids ran over to have a look at her, it really was amazing, it had everything on it. It was making a road as it went along.' *[slap slap slap]* 'It crushed the trees down.' *[slap slap slap]* 'It was following the railway line crushing the timber that was in its way on each side as it came into line.'

Julia was describing a mechanical beast known as Big Lizzie. Built during World War I, Big Lizzie was a slow-moving giant tractor that famously ate its way through the Australian bush. Rightly described as 'the mother of all tractors', she was 34 feet (10.4 metres) long, 18 feet (5.5 metres) high and weighed 45 tons (46 tonnes). With her two trailers attached she stretched as long as 98 feet (30 metres). The grunt came from a 60-horsepower crude oil engine that could only push her to a maximum speed of two miles per hour. If she wanted to turn around she needed a radius of 200 feet (61 metres). The slapping noise that Julia remembered so vividly came from the dreadnought wheel design. That meant Big Lizzie travelled on her own in-built tracks that were fitted to each wheel and slapped into place as the vehicle crawled along. The designer, Frank Bottrill, wanted a vehicle capable of travelling over soft sand. What he created was the ultimate weapon in the battle against the bush.

Big Lizzie left Melbourne in 1920 and crawled to Red Cliffs, on the Murray River, just inside the Victorian border. For five years she chewed through the native bush, tearing out trees, including mallee—from the soft sandy soil. The war was over and soldiers were coming home to a bright future—the new battle was for development and Big Lizzie was the heavy artillery. Some bush was burned, sometimes chains were used, and then picks, crowbars and axes cleaned things up. The Commonwealth government selected and acquired sites for returning soldiers and then the state governments granted the land. In Victoria, the mallee was a particular target. This rough bush needed to be 'developed'— instead of eucalypt forests stretching from horizon to horizon, they wanted fruit blocks and vineyards. River red gums were chopped down for fencing, wharves and as fuel for paddle-steamers. Logs kept in stacks in South Australia generally weighed about 100 000 tonnes. The work was exhausting and life was basic—photos from the era show tiny houses made of fibrolite (asbestos cement sheeting).

The vegetation that had evolved over 30 million years was removed in half a decade, and in 1925 Big Lizzie was silenced. In her wake was the largest soldier settlement in Australia.

Big Lizzie might have disappeared into legend had she not been discovered in the 1960s rusting away on a property in western Victoria. A campaign raised enough money to buy her and she made the return trip to Red Cliffs, in time for the town's golden jubilee in 1971. She remains in pride of place in the town as a tourist attraction.

Red Cliffs is not a cowboy town—far from it. It is part of a region that produces 40 per cent of Australia's food. Wander through a supermarket one day and look at all the Australian produce and products. Then consider four out of every ten of them are from this region and you start to understand the value of the Murray–Darling Basin food production. Food production in Australia is estimated by the federal Department of Agriculture to be worth $49 billion. That industry started in places like Red Cliffs with returned soldiers clearing small blocks of bush and putting in fruit trees.

After clearing had finished an estimated 15 billion trees had been felled in the basin. This is the extraordinary figure quoted by the Murray–Darling Basin Commission Chief Executive, Don Blackmore. This seemingly impossible task shows the determination and incredible commitment white Australians have had to turning this unusual and fragile country into a southern hemisphere version of Europe.

RIDING THE SHEEP'S BACK

Big Lizzie was the culmination of a process that had begun in 1797 when a supply ship carrying 10 Spanish merino sheep pulled into Sydney Cove. These remaining sheep of the 26 that had originally left Britain represented the future of the colony—agriculture. Australia's first farming systems, like their livestock, were all imported from Europe. By the 1870s wool was the dominant export from the colonies and, along with wheat and gold, represented the basis of the Australian economy for almost a century. By the end of the nineteenth century New South Wales alone had more than 60 million sheep. These animals with hard hooves and ferocious appetites needed large pastures in the new country. The land with its poor soil and unpredictable rainfall couldn't support the large flocks in the same ratio as the northern hemisphere. Land had to be cleared and stations of legendary proportions were created. The determination to survive droughts and floods and the vagaries of world markets has left an industry that remains worth $3 billion a year in exports. The generations of sheep that followed those original 10 merinos now produce 600 million kilograms of shorn wool per year.

Pastoralists searching for land were soon in competition with farmers looking for grain fields. While sheep farming was developing a squattocracy, improvements in farming methods in the 1860s and 1870s created better opportunities for growing crops. The desire for land was now pushing land clearing into the most difficult of country to clean up—the mallee. Again local engineering came to the rescue in overcoming the shortfalls of traditional techniques. A mallee or scrub roller was invented which a horse could pull. It was a simple enough process: everything the roller could get over the top of was flattened and the mallee trees ignored. Then once everything had dried out, fire was set to the place. Once the fire burned itself out (which must have been quite a wait given the density of wood growing in mallee forests) the land was raked clean. The dead trees were cut down and a wheatfield planted. The only problem was they didn't realise the mallee's great survival organ buried beneath the ground hadn't died. With its natural adaptation to fire, the mallee would reshoot and have to be subdued again. In addition, the roots would constantly catch and trip up the plough being dragged behind the horse. When you consider what level of determination a farmer working such conditions would have needed, it's no surprise that the man who invented a remedy was given a medal.

In 1876, Richard Smith was just another farmer breaking his back trying to grow wheat in mallee country at Kalkabury on South Australia's Yorke Peninsula. He figured there had to be a way to beat the stumps that tangled his fields. With the help of his brother Clarence, who was a blacksmith, they produced a piece of machinery they called 'Vixen' but which came to be better known as the 'stump-jump plough'. Through a series of levers and springs, the plough would literally jump out of the ground before hitting a stump, leapfrog the problem and return to the earth to continue ploughing. So delirious was the South Australian Colonial government they gave Smith the medal, another 640 acres of land and £500 to play with. Within two years of the stump-jump plough being invented the push for wheat reached a new level. Between 1878 and 1881 more than half-a-million acres of land was cleared to grow the grain.

TWO CANADIANS AND A DREAM

Big Lizzie isn't the only piece of machinery that has become iconic in the Sunraysia region of northwest Victoria. In Mildura, a city sometimes referred to as the 'oasis of the mallee' there is a place called Psyche Bend. There, sitting in state, is an engineering marvel of similar proportions to the great scrub clearer. Its full title is 'The Chaffey Triple Expansion Steam Engine and Tangye Centrifugal Pumps' but it could easily be known as the first-ever irrigation system used in Australia.

By the late 1880s the stump-jump plough and merino sheep had created enormous agricultural wealth in Australia. However in the Victorian Parliament sat a young man who dreamed of a future for the

Victorian colony that went beyond wool and wheat. Alfred Deakin was only 27 years old but already a Member of Parliament and Solicitor-General of the Liberal–Conservative Coalition government. He was also the Commissioner of Public Works and Minister for Water Supply, a task he threw himself into. Deakin was a brilliant thinker and a man who liked to entertain his mind on a grand scale. He became fascinated with water and irrigation. While the young minister made a name for himself by bringing social reforms to industry and halting child labour, his thinking constantly turned to the north of the state where the mighty Murray River formed a border with New South Wales. To search for ideas on how to use this resource, he read reports about water systems across the world and in 1886 travelled to the United States on a fact-finding mission. What he found were two Canadians who seemed to understand what he had in mind and shared his dream for the future.

The Chaffey brothers were running a successful irrigation business in California at the time of Deakin's visit. They were brilliant engineers who had worked with their father in the construction business in Canada. The elder son, George, left his father's business at the age of 30 when he took up the post of chief engineer of the Los Angeles Electric Company. A clever man, he was responsible for making the city the first all-electric one in the world. After this remarkable feat, he left the company and turned his attention to the deserts of California and their potential for irrigation farming. His brother William joined him and they started a fruit farming community by turning 8000 acres (324 hectares) of desert into orchards. Although well established they seemed ready for another challenge. Deakin offered them carte blanche if they agreed to come to Australia to set up an irrigation colony on the Murray River. The brothers agreed to have a look at a location called Mildura Station (later to become the site of the city of Mildura) and were impressed enough to sign on for the project in October 1886. However things didn't quite go to plan for Deakin with the visitors causing some insecurity among Victorians already in the region and their political representatives in Melbourne. The Victorian Parliament stalled the plan with debate and in one of the earliest

VIEWPOINT
Ian Donges

IAN DONGES FARMS CEREAL, LAMBS AND BEEF CATTLE AT HIS
PROPERTY NEAR COWRA IN NEW SOUTH WALES. HE HAS BEEN
INVOLVED IN FARMING ORGANISATIONS SINCE 1978, SERVING ON THE
SHEEPMEATS COUNCIL OF AUSTRALIA, THE GRAINS COUNCIL OF
AUSTRALIA AND AS THE PRESIDENT OF THE NATIONAL FARMERS
FEDERATION FROM 1998 TO 2002.

> *I am sure I speak for many farmers in Australia
> when I say the thing that I care about most is the
> quality and the productivity of the land. So when you see
> salinity coming into a property you just see something that is
> the opposite of what you want. The key is what can you do
> about it? That's what is so frustrating for many farmers
> because the cause of their problem may not be anywhere near
> their farm.*
>
> *What we have seen in the last five or 10 years is an
> increase in outbreaks of salinity around properties where it
> was once never considered an issue. Now we're finding out the
> cause may have been due to things we did 40 or 50 years ago.*
>
> *We always learn from history and when we look at the way
> farms were created in Australia we have learned a lot. We*

originally picked up English farming methods and went round and round with ploughs and things that were obviously a severe disadvantage to our soil. We've learned from that certainly as far as tillage is concerned. We've learned that we've taken too much vegetation from some landscapes, which has impacted on some watertables, which has impacted on the salinity.

The trick is now how are we going to use all that knowledge and get on top of the problem? I think we've got time but I think we've got to work at it through community and government and a lot has to be done in a short period of time.

Salinity is one of the major threats that we have to Australian agriculture. When you talk about perhaps 10 or 15 million hectares being affected in 20, 30 or 40 years time and some of our major rivers becoming too saline to either use for agriculture or for drinking water, that says a lot to me about the serious impact that could have on all of us.

Quite obviously it is a whole-of-community issue. Yes the farming industry has a strong vested interest in it because we own, or are responsible for, 70 per cent of the Australian landscape. The farmers of this country have already demonstrated they are very willing to participate in any rectification or control mechanisms that we can come up with. In my community there are a lot of very good innovative young farmers who are changing their farming systems by using minimum tillage and alternating crops across their land. That's very positive. But I think the biggest solutions are still to come in terms of the whole-of-community approach.

I wouldn't be a farmer if I weren't an optimist. I am very optimistic about the future and I am very optimistic about our ability to get on top of this problem.

examples of what would become an ongoing interstate rivalry, South Australia swooped on the Chaffeys. Led by Premier John Downer, the neighbouring colony offered the Canadians a similar site, at what would be known as Renmark less than 150 kilometres downstream from Mildura. The Chaffeys accepted the site and on 14 February 1887 they signed on to create the irrigation colony of Renmark. The Canadians worked on 30 000 acres (12 150 hectares) and planted vineyards and orchards that were irrigated by a system of open drains. George started his own farm on 160 acres (65 hectares) at Paringa, near the South Australian border, where he planted groves of citrus and olives, vineyards and orchards of stone fruit. As well, he added windbreaks of native trees, planted wheat and lucerne for his dairy stock and built a Canadian-style log cabin to house his wife and the first of his six children.

Within three months of work beginning across the border, the Victorian Parliament passed the Chaffey brothers agreement. In August 1887, Alfred Deakin's dream became the Chaffeys' 250 000-acre (101 250-hectare) reality. Chaffey Brothers Ltd was more than just an irrigation company. It influenced everything including the layout of Mildura, which is similar in design to their hometowns in California and Canada. The town's centrepiece is a boulevard that used to run through the plantation and be serviced with electric trams. This they named Deakin Avenue after their great benefactor. The brothers began work on stage one of the project that involved converting 50 000 acres (20 235 hectares) of former mallee scrub into orchards and vineyards. It was a modest start: their deal with Deakin involved a commitment to irrigate over ten times that amount of country. There was never a question that the liquid gold of the Murray River was an unlimited resource. Progress had already felled a majority of the great river red gum forests along the river.

Although the brothers found the desert-like mallee country less fertile than California, they were sure that if they added enough water, eventually the red sand would yield bountiful crops. As William worked on the horticulture, George designed the great machine that would drive the water out of the riverbed and into the orchards. It was built in Birmingham, England, and her boilers drove a four-cylinder engine rated at 1000 horsepower (746 watts). Four years after arriving in Mildura, the Chaffeys unveiled the Psyche Bend Pump Station Engine, a system capable of pumping 40 000 gallons (182 000 litres) of water per minute. In 1891 it thumped into life for the first time and kept pumping until 1959.

The Chaffeys' water licence allowed them to pump the equivalent of 600 millimetres of rainfall onto the land every year. This country when it was covered in thirsty mallee trees normally received about one-third of that amount. With those trees replaced by shallow-rooted vines and fruit trees, the sandy soil was absorbing one hundred times as much water under irrigation as it had been in its natural state. The oranges, grapes and apricots were growing but so was a silent flood beneath the ground.

Despite their influence, the Chaffeys didn't both stay in Australia. The irrigation colony struggled during the depression of the 1890s and the Canadians drew much of the blame. George left his bankrupt business for California, leaving behind his homestead 'Olivewood' at Paringa. William stayed on though and rebuilt his business and Mildura. He even became mayor in the 1920s and is lauded with a large statue outside the city offices where a plaque honours his dedication in creating both irrigation and the dried-fruit industry.

Irrigation in Australia has boomed during the last century and is a major reason why the Murray–Darling Basin is the country's food bowl. George Chaffey's great pump has been replicated in thousands of different forms on every one of the 24 major rivers in the Murray–Darling Basin. The ethos of the era, that land clearing and then irrigation was an imperative for the nation's future (and possibly even a sin against God not to take up), is well illustrated in a pamphlet from 1917 advertising the potential of the Loxton district in South Australia:

> The Murray has always been a runaway. Only now in South Australia, with the other states, [are we] beginning to put the harness on it, at a cost of many millions of pounds. For countless years the greatest river in Australia has been doing practically nothing more than empty its fertilising volume into the sea. River mathematicians have calculated what quantity of water has been wasted in that swift, unceasing journey of gold to the ocean; but statistics do not matter. Not what was, but what is, matters. It is true, however, that the Murray has been damned by the politicians rather than by the engineers. The dawn of a new and happier era has broken on the horizon of Australian statesmanship.

Irrigation and dryland farming had its next major surge thanks not to the agriculture of Europe and America, but to the aftermath of war.

BRINGING THEM HOME

Once started, land clearing continued unabated in every state. In Queensland and the Northern Territory the rate of clearing was slower because the nature of the tropical climate and vegetation meant that land clearing wasn't always an option. The second great wave of scrub clearing came when a grateful nation sought to reward the hundreds of thousands of servicemen who had fought during two world wars. Thousands of men were given blocks of land that mostly had already been cleared for them. After World War I this was a strenuous task. There was only one Big Lizzie and she was kept busy for a long time near Red Cliffs. By the time World War II ended, Big Lizzie had given way to a new generation of diesel-powered iron bullocks that could splinter forests in times previously unseen.

A propaganda film of the era is now tucked away in the national screen and sound archives at ScreenSound Australia. Entitled *The Farmer was a Fighting Man* it features scenes from Kangaroo Island, Mildura, Tasmania and Western Australia. There are images of tractors pulling chains and heavy balls through the bush and at one point a man in a suit almost leaping for joy like Peter Pan with a torch in one hand setting fire to the scrub. In the traditional newsreel patriotic style of the times, an announcer with a toffy accent boasts of the great achievements being made in the soldier–settler schemes:

> For those looking to the land the War Service Land Settlement Scheme was planned and put into operation in various states.
>
> South of Perth in Western Australia hundreds of thousands of acres of good soil with adequate rainfall was available.
>
> These areas have so far been left unimproved owing to the heavy cost of clearing. So where the axe has failed, settlement scheme mechanism gets going in a big way.
>
> This is a vast national undertaking and like the building of bridges and the making of dams it is on a scale that can only be carried out by government.

Tractors and bulldozers, the draughthorses of today are used in a hundred different ways … they can do in hours what would take weeks of human labour.

So this huge investment will pay rich dividends in the future and with this scheme Australia adds to her achievements as one of the great food producing nations of the world.

Western Australia didn't get going as fast as the eastern states. However when the first push came in the 1920s and 1930s and then again in the 1950s, the state made up for lost time. History suggests the west always had more of a salt load than anywhere else in the country. The brilliance of the farmers in clearing and growing food and fibre eventually was to create the largest salinity problem in the country.

Dr Tom Hatton grew up in California, a place also known both for its agriculture and emerging salinity. He worked as a hydrologist in the Murray–Darling Basin for a decade before moving with the CSIRO Land and Water Division to Western Australia. He talks about the land clearing in Western Australia in the early part of the twentieth century with a sense of awe: 'I honestly can't imagine it. The early stuff would have been done by hand with horses and then post war it was mechanised but they were clearing a million hectares a year even in my lifetime.' He pauses and shakes his head in admiration. 'That's a lot of work and it established a lot of good farms and a lot of food for people.'

Wheat farming in Western Australia drives the Australian grain industry. Caught in the evening light, a crop ready for harvesting glows as if lit from a heavenly source. When the breeze catches it, it ripples and sways like an amber sea. As the third largest exporter of wheat in the world, Australia provides 17 per cent of the world's needs, which earns an annual return to the nation of almost $4 billion.

Wheat farming started later in Western Australia than elsewhere. Initially Tasmania was the nation's granary but South Australia took the lead in the nineteenth century with the invention in the 1840s of the stripper that mechanised harvesting and made the industry more efficient. Soon after, the crop was also taken up across the Yorke Peninsula, and what we now know as the Wimmera and Mallee regions of Victoria and the Riverina region of New South Wales. By the turn of the century New South Wales was the largest grower, but at the same time the Western Australians began clearing land in the southwest of the state and by the 1960s the west was awash with wheat. Richard George from the Western Australian Department of Agriculture says it was the railway that opened up the country to wheat farming: 'There was development near Perth around the turn of the century but it really only got going in the 1920s and 1930s when the railway lines went out through the wheatbelt and farmers got into clearing the heavy country, as they called it then. It was a massive job clearing this country. I can hardly imagine what it was like to do that. It obviously took a long time and it was probably really only after World War II when heavy machinery became available and when trace elements became available that clearing of a lot of the upland took place. Clearing of the wheatbelt was a staggered process with a couple of bursts in the 1920s and 1930s and then 1950s and 1960s.'

While walking around a degraded patch of saline land southeast of Perth, Richard George confirms the ancient history of the western seaboard is the same as the east. Salt was deposited with wind and rain and because of low gradients took a long time to drain out. The thirsty trees kept the watertable low and so the system was in equilibrium. Like Tom Hatton, Richard George respects the early farmers and says they were driven hard to clear their land.

'There was a very strong development ethos. We had a growing population and we obviously needed food to support the growing colony as it was then. Agriculture was in development right across the world. Dryland agriculture is what we practice in Western Australia and the colony needed development to progress from its fledgling status. Gold and wheat and wool formed the basis of that development.

'It was policy to clear. Your holding was conditional on clearing and in many cases you would have lost your land had you not

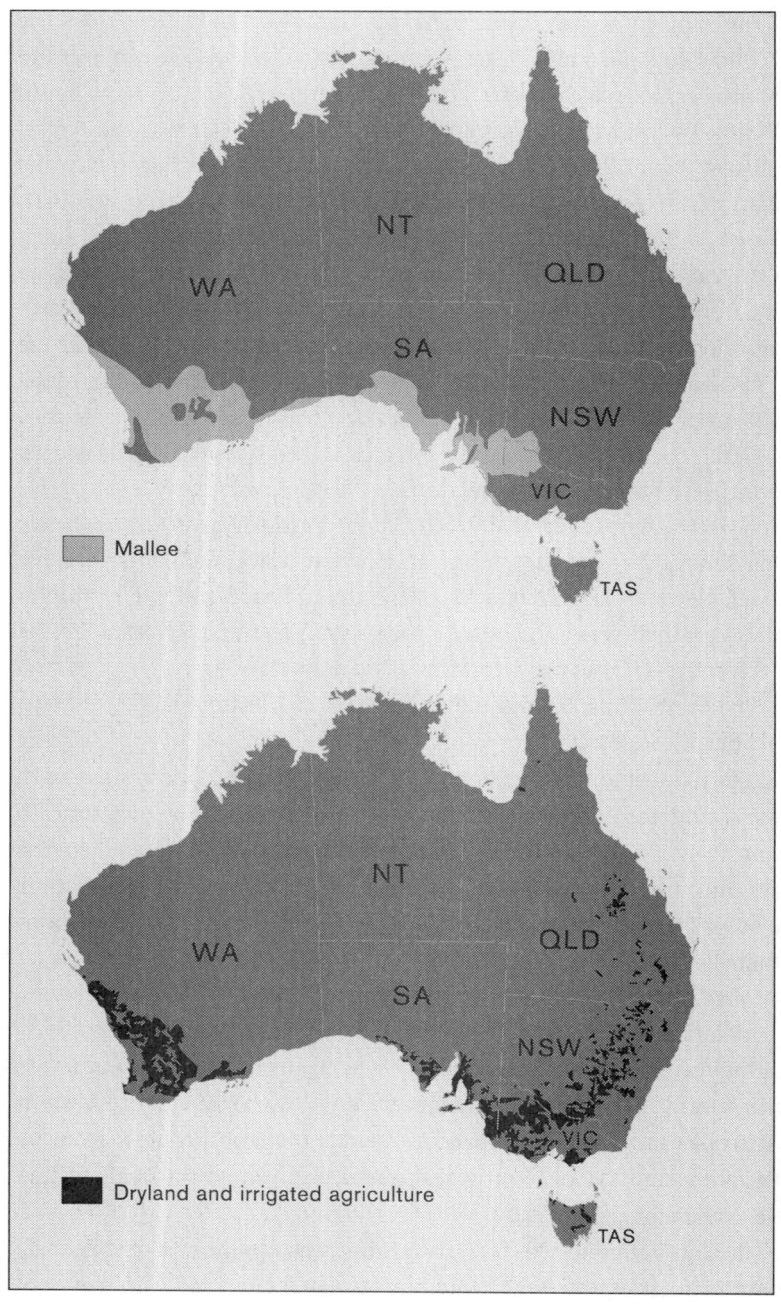

Mallee

Dryland and irrigated agriculture

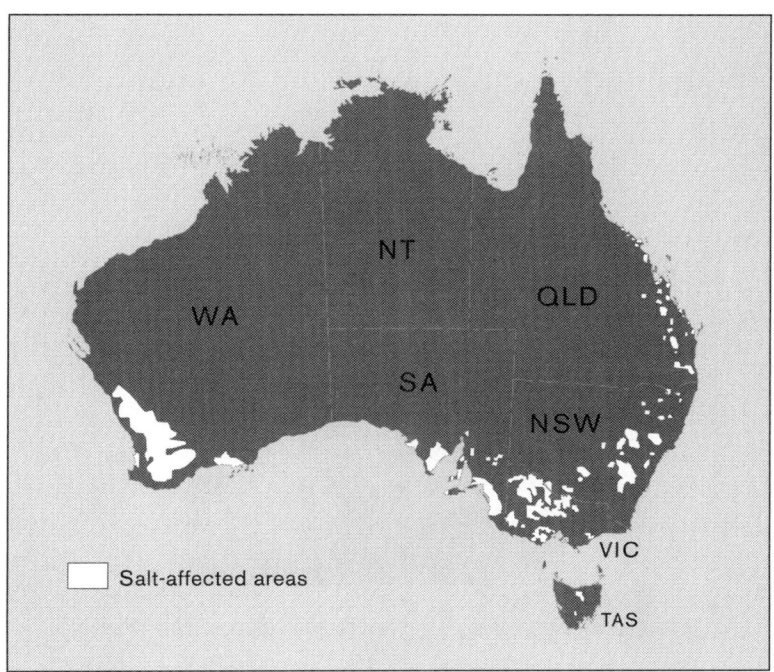

Changes to the Australian landscape leading to salinisation. The mallee map (top left) shows the original coverage of mallee forest before these areas were cleared for farming. The dryland and irrigated agriculture map (bottom left) shows the intense agriculture that followed and almost mirrors the extent of salinity in the salt-affected areas map (top right).

followed the conditions of your land use. The farmers were basically acting on the conditions of their land ownership.' (*See also box, Richard George p44–45.*)

Ian Stanley from Kalannie, north of Perth, is typical of farmers of the wheatbelt. It was his grandfather who started clearing the land. The family photo album shows grinning family members standing by tractors and holding up signs proclaiming record yields. It was, and still is, an extremely successful merino sheep and wheat farm. It also has salinity problems. When commenting on the general cause of these problems, Ian says, 'There are probably areas that shouldn't have been cleared, but that's easy in hindsight.'

VIEWPOINT
Dr Richard George

RICHARD GEORGE RUNS SALINITY PROGRAMS FOR THE WESTERN AUSTRALIAN DEPARTMENT OF AGRICULTURE. HE HAS A PhD FROM THE UNIVERSITY OF WESTERN AUSTRALIA AND IS AN ASSOCIATE PROFESSOR AT MURDOCH UNIVERSITY. HE SPECIALISES IN HYDROGEOLOGY AND IS ON THE BOARD OF THE NATIONAL DRYLAND SALINITY PROGRAM.

Western Australia has the largest salt stores in Australia. We have very strong winter rainfall and strong evaporation in summer. We also have very low gradients so the salt accumulates.

Salt has been accumulating in WA at a rate of about 20 kilograms per hectare for probably 30 to 100 000 years.

In the wheatbelt of WA we have probably got up to 5000 to maybe 10 000 tonnes of salt stored under every hectare. Essentially, prior to clearing what came in went out. It was an equilibrium: salt coming in the rainfall and going out with the runoff. Now, there are a couple of Sydney Harbours' worth of water going under the wheatbelt every year. We are halfway or a third of the way down the track of salinisation. We are probably not going to do much to the next third unless we get into engineering in a big way.

What I want to reinforce is that there is the other 70 per cent of the landscape which requires maintenance and protection because it's going to carry the can if we fail with this path of salinity management. So we've got to make sure as a country that we continue to invest in salinity management but we also invest in the protection of the rest of the agricultural landscape.

We understand the process of salinity. That's nothing really new. To actually address salinity to make a difference in the way the landscape is going to look we have got to get economic systems in place that farmers can afford to adopt at the moment. Some of them are research at best and others are just not practical.

Our rural history is deeply embedded in Australia's national identity although farmers represent less than 1 per cent of the population. The development ethos that encourages land clearing has continued throughout the twentieth century and now with the use of irrigation even crops such as rice and cotton can be produced. Most of what is grown in Australia heads overseas. According to the Department of Foreign Affairs and Trade, in 2000–2001 exports of primary produce earned Australia $70 billion. It was not until the 1980s that federal subsidies for clearing were finally abolished by the Hawke government.

Australia has ridden the sheep's back ... but at what price?

CHAPTER 3
The Salt Harvest

Patter, patter ... Boolcoomatta,
Adelaide and Oodnadatta,
Pepegonna, parched and dry,
Laugh beneath a dripping sky.

C.J. DENNIS, 'A SONG OF RAIN'

Australia is a nation that enjoys irony. Originally it developed as a way of coping with this harsh country. Australian cartooning is full of scenes of men in dusty godforsaken locations making ironic wisecracks out of the corner of their mouths. However there is one great environmental irony that gets very few laughs. In a country that is the driest inhabited continent on earth, salinity has developed because of too much water. While across the nation from farmers to environmentalists to urban dwellers we are obsessed with the need for rain, salinity is being exacerbated because too much water is getting down into the earth.

John Williams from CSIRO, explains: 'Most of the restraints on our agriculture, in terms of productivity, are related to having not enough water and not enough nutrient. Yet the problem that causes salinisation and much of our land degradation is an excess of water beneath the root zone and an excess of nutrient below the root zone at the wrong time of year.

'Many of the problems of salinity are also associated with waterlogging. So at one time of the year our wheat crop is waterlogged and yield is being suppressed because of that, and another time of the year the yield is being suppressed because there isn't sufficient water. This is because our agriculture at the moment is hydrologically out of balance with the way the Australian landscape works.'

The land-clearing ventures of the nineteenth and twentieth centuries have left the country barren and exposed to rain. The great forests are gone and so is their capacity to soak up rainwater. So much water now gets into the Australian watertable that the country is filling up. While above the ground our land may be parched and decaying from lack of water, below it the water supply is increasing. It is probably the most bitter of Australian ironies but one that we've known about for more than a century.

THE FIRST SIGNS

In 1924 Walter Wood published a paper in the Journal of the Royal Society of Western Australia with the apt, if not blunt, title 'Increase in Salt in Soil and Streams following the Destruction of Native Vegetation'. In a world where theories come and go, Wood's historical paper contains a classic description of dryland salinity and one that is still considered to be mostly correct.

Wood was a bushman who had grown up in various parts of South Australia, including the Yorke Peninsula, about the time the Smith brothers were being hailed as heroes for inventing the stump-jump plough. He had a sharp mind and keen eyes and worked as a surveyor and engineer in his home state before moving to Victoria and later Western Australia where he worked for the railways.

Concerns about salty soils in Western Australia were already real enough to cause a Royal Commission in 1917 into the development of mallee land at Esperance. Despite evidence that one-third of the land was too salty for farming, development was recommended. The spreading railway needed support systems of which the most important was water for the boilers. As early as 1905 there were reports that the water was becoming too salty to be effectively used by the trains. Apart from the corrosion it causes, salty water has a higher boiling point and so is less efficient for steam engines. Eventually some water was so brackish it was unusable and an investigation was ordered. The problems were initially overcome by engineering but Wood pursued the salt back to its source. His bush logic, scientific approach and sheer doggedness led him to follow up a theory he had heard in his youth, as he explained in his 1924 paper:

> For many years I have been interested in the fact that in certain districts in the southern portion of Australia where destruction of native vegetation has taken place rapidly, there has followed a very noticeable increase in salinity. I first noticed this over 30 years ago in Yorke Peninsula, South Australia.

Wood tracked down other theorists who'd noted the connection between land that had been heavily cleared and saline groundwater. In 1917 a railway engineer, Robert Bleazby, wrote a paper entitled 'Railway Water Supplies in Western Australia: difficulties caused by salt in soil'. It reported that salinity had been overcome by revegetating some sections of a catchment. Wood believed that following clearing, water was able to penetrate the earth and absorb the salt that sat in layers beneath. He also theorised for the first time that the salt was brought in by wind and rain and not by marine sediments. This last idea earned him some derision at the Royal Society of Western Australia and his reports had little impact on the development ethos of the state. Wood might have drawn some snorts of disrespect that day but, close to a hundred years after his paper was published, an award was established in the name of Walter Ernest Wood for those scientists whose work adds most to the understanding of salinity in Australia.

One man who has won the Wood award for research into salinity is Richard George from the Western Australian Department of Agriculture. He believes Wood began putting two and two together on salinity as early as 1897, and so by 1924 was very confident his work was more than a theory. 'His view of the hydrology of the wheatbelt is fairly much the same as now. That excess clearing caused water to go into the watertable and that as that watertable rose it dissolved salt in the subsoils. That salt slowly came out with groundwater and ran to accumulate, in his case, into the dams used for water for railways.'

Tom Hatton from CSIRO Land and Water Division, has also won the Wood award and sighs a little when asked about the link between land clearing and salinisation: 'That landmark scientific paper that established [a link] beyond all reasonable doubt, was published in 1924 so for more than 75 years we have known what causes salinity. Whether we have heeded that advice is another question.'

IT'S A GROUNDWATER PROBLEM

When scientists are asked to explain how salinity works they often furrow their brows and say, 'It's not simple'. Often it isn't because salinity levels are determined by soil type, rainfall, sunlight, gradients, catchments, tree species, farming practices, leaking pipes, building codes and irrigation—and last but not least, what the neighbours are doing. There is, however, a simple explanation to the most basic question of how salinity works in urban or rural settings. As Oxford University hydrologist Warren W. Wood says, 'Basically it's a groundwater problem.'

Every time rain falls and trickles down through the soil profile and adds to the watertable, it starts to build and rise. The rate at which watertables rise is naturally dependent on the rate of water coming down from above. If you have a thirsty forest at work and only a teacup full of water coming through every year then it would change very little. However, if most of the rainfall is moving straight through the system, the watertable will build at a brisk rate. If in extreme circumstances the watertable keeps rising, it will eventually fill every part of the earth that previously held air. The ground then becomes waterlogged because the watertable has actually reached the surface and flowed out onto the landscape. In waterlogged areas the lower parts of the landscape will become submerged as the watertable continues to rise. That is the first step towards dryland salinity.

The second step is the salinisation of the water while it sits underground. Water reacts with its surroundings. If there is a rock casing (aquifer) for it, it will remain almost as pure as it was the moment it fell from the sky. Australia has one of the world's greatest underground rainwater tanks. It sits under about one-fifth of the continent and is known as the Great Artesian Basin. Beneath the earth, between layers of sandstone, is the equivalent of 17 000 Sydney Harbours of water. The basin began forming more than 200 million years ago and some of the water is believed to be up to 2 million years old. At times there has been enough upward pressure for springs to burst forth in outback areas. These artesian springs are known as 'mound springs' and within a morning's drive from Lake Eyre, you come across one of Australia's most famous, Dalhousie Springs, where clear, sweet water pours into the desert 24 hours a day. Flocks of birds wade through the wetlands while local ringers wash off a few days' worth of dust in the natural whirlpools.

For more than a century, graziers and other developers have drilled down into the basin and used the water not only for irrigating stock, but also for running turbine and steam engines. The original Ghan railway line from Port Augusta to Alice Springs was able to use the water to slake the thirst of the iron horse as it puffed its way through the hostile outback. Today WMC Resources (the former Western Mining Corporation) uses basin-water to run its giant copper, gold and uranium mining operation and support facilities at Roxby Downs in South Australia in an area with an average rainfall of less than 200 millimetres. Some parts of the basin have saline water but mostly it remains pure. The Great Artesian Basin is still recharged each year with rainfall that seeps through sandstone at the back of the Great Dividing Range but the watertable is dropping due to the amount of water being extracted.

VIEWPOINT
Dr Warren Wood

WARREN WOOD RECEIVED HIS PHD FROM MICHIGAN STATE
UNIVERSITY. HE IS CURRENTLY A RESEARCH HYDROLOGIST WITH THE
US GEOLOGICAL SURVEY, A VISITING LECTURER IN THE DEPARTMENT
OF GEOGRAPHY AND CLIMATE CHANGE AT OXFORD UNIVERSITY AND
EDITOR-IN-CHIEF OF THE TECHNICAL JOURNAL, *GROUND WATER*.

*Basically salinity is a groundwater problem. It's
the result of rising watertables. All water has
undissolved salts in it. If you raise the watertable up to the
point where the capillary action draws the water up
evaporation off the surface occurs and concentrates the salts.
The salts themselves come from the ocean: in other words the
rainwater has a certain amount of salt in it, and also some
comes from the dissolution of the subsurface rocks themselves.*

*There is not much you can do about the amount of salinity
brought in by precipitation—all you can control is the
groundwater level. You can do that by extraction using wells*

or you can do it by vegetation that 'pumps' the water out by using the water. The trick is to keep the watertable at least a metre or 2 metres below the surface level. But this also depends on the soil type. With very fine soil you may need more than 2 metres.

You only have a few milligrams of salt per litre of rain but if you do the simple calculations you see that over a period of hundreds of thousands of years, the salt accumulates to thousands of kilograms per square metre. Unless there is a flushing action to remove this salt then it continues to build up.

There are two factors involved with groundwater flushing. The first is hydraulic conductivity, or the ability of water to move through the rocks. For example, water flows much more easily through gravel than through sand or clays. The second is the slope, and in the case of Australia, the slope is very low. You don't have huge topography to drive this system so as a consequence, the amount of flushing is relatively small and it's very easy to build up significant amounts of salts over time.

The river system and the groundwater system are intimately connected. The two are exactly the same thing, just a different face of the same picture. All groundwater discharges to streams so ultimately if you have salinity in the groundwater you are going to have salinity in the streams. If you irrigate with this water you are going to exacerbate the problem. There are two aspects to salinity—one is due to irrigation, the other is due to dry land.

Other subterranean water supplies are not so lucky. Far from the sandstone that keeps the Great Artesian Basin water in such good condition, other groundwater sits in old, tired soil that has been accumulating tonnes of salt for thousands of years. As the water sits underground it absorbs the salts causing the groundwater to turn saline. When saline groundwater builds and the watertable rises, it continues to absorb salt on the way up.

The third step in the hydrology of salinity is the role of the sun. The combination of heat and wind dries out moisture from the surface. Everything, from lakes to tree leaves, has moisture pulled out of it. In the case of groundwater, the evaporation process can begin when the watertable is still as deep as 2 metres below the surface. This is caused by what is known as capillary rise, and it

MEASURING THE WATERTABLE

Whenever scientists, engineers or farmers approach a saline area the first question they tend to ask is, 'Where's the watertable?' If the answer is 'Don't know' then it is time to put in a piezometer. Piezometers are probably the most simple and essential measuring tool for salinity because they measure the water pressure or compressibility under the earth.

Scientists seem to agree that once the watertable reaches within 2 metres of the surface then the capillary action begins. This means that the sun is pulling moisture out of the topsoil and in turn this draws up the water from a depth of 2 metres, bringing with it the dissolved salt.

Inserting a piezometer is done by digging a non-pumping bore, usually about 3 metres deep—there is no point going any deeper than that because if you haven't struck water by then there probably isn't much of a salinity risk. A piezometer can be as simple as a small plastic pipe, roughly the size of a fluorescent light tube, tunnelled down into the ground. Into this an even thinner, but longer, plastic tube is inserted with a float on the works in basically the same way a dry sponge does when placed onto water spilled on a kitchen floor. Eventually the water will rise up through the dry sponge as the moisture moves into the dry spaces. As the elements dry out the surface soil, the water below is attracted upward and in turn evaporates, starting the process over again.

While this process reduces the level of the watertable, the trouble is the salt that has been absorbed by the water is taken up when the water evaporates. It makes it to the surface with the water and then stays behind, creating a salt harvest.

Even while driving his car through the Western Australian wheatbelt, Richard George can spot the first subtle signs of salinity. He stops the

---

bottom. The float will eventually find the watertable and float on top of it. The landowner can then keep a check on where the level is by seeing if the inner pipe (the top part of which sticks out above the ground) has risen or fallen. Alternatively, just an outer pipe is used with a cap over the top. A measurement is made by lifting off the cap and lowering a length of string with a weight on the end down into the pipe until it hits water. Then the string is brought out and measured.

Farms with rising watertables may have many piezometers spread around the property to measure where watertables are rising and falling. By attaching bright plastic flags to the top of the measuring tube a quick visual reading can be made from the back of a tractor.

Piezometers serve the same purpose in cities and towns. Homeowners concerned about salt affecting their houses can use one to find out where the watertable is and whether they need to begin taking potential measures against salt damage. Wagga Wagga, in New South Wales, has installed close to one hundred piezometers in the city to keep a very close eye on where the watertable lies.

vehicle, gets out and scratches a piece of ground that is showing tinges of a white crust. 'Salt scald,' he grunts as he bends down to scoop up a handful of salt crystals that have accumulated on the surface. Nothing grows near them and the earth around is barren and wounded. 'Land is normally considered salty when it's got enough salt in the root zone to affect the production of the particular crop you are going to produce. There are various thresholds for various crops. In the wheatbelt there are a lot of visual indicators as well. Some signs are when the land gets a covering of sea barley grass and you see salt at the surface effervescing. Often it is subsoil salt and salinity in combination and it really depends on the type of crop you've got. But normally in the wheatbelt, the signs are a loss of productive cover. We tend to refer to salinity developing once you have lost between 10 and 50 per cent of your crop growth.'

Richard explains that the process of creating dryland salinity may take decades and while it is brewing underground there is no sign of the danger ahead. Once it reaches the surface in small doses, it is only a matter of time until the salt takes over. 'A shallow watertable and the evaporation off that table is the cause of salinity. But it might take a few years before the salt builds up to a concentration that will kill the particular crop that you have got growing. Rainfall, leaching and also soil variability mean that your paddock may not become as salty as your neighbours'. But it just takes time—five years or so once the watertable comes close to the surface.'

A study by the Western Australian Department of Agriculture in Perth compared land that had been cleared with land that retained its native vegetation. It showed that while annual evapotranspiration and salt input rates remained roughly the same, the salt output from the land went from 5 grams per square metre in covered country to 500 grams once the vegetation was gone. With deforestation our land is getting saltier—and so is the extent of dryland salinity.

IN THE RIVER

Central Australia has the most outstanding collection of salt lakes but nowhere is dryland salinity more obvious than in Western

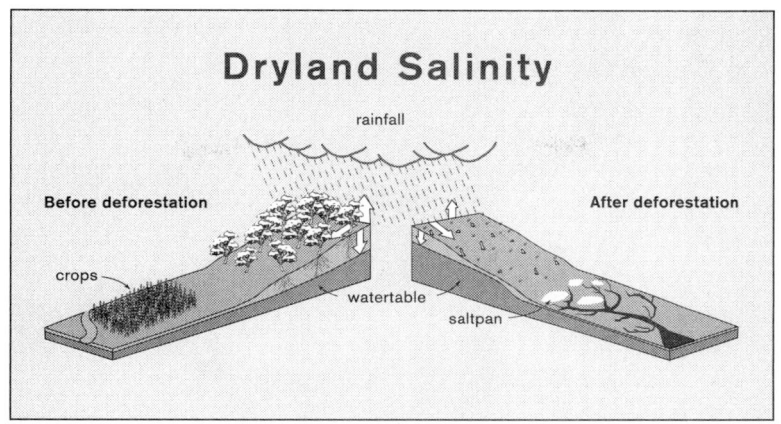

Dryland Salinity

rainfall

Before deforestation

After deforestation

crops

watertable

saltpan

Dryland Salinity is caused by deforestation; as the watertable rises is brings salt to the surface.

Australia. Even an uneducated observer glancing out the window of a passenger jet can't help but see the brownish-grey rings that mark the earth below. As the extent of dryland salinity increases so does the harvest of salt that now flows through almost every creek and river in the area. Freshwater salinity may seem a contradiction in terms but a by-product of dryland salinity is the salty creep into the waterways of Australia. Like when Charles Sturt tasted the water of the Darling River in New South Wales, Western Australian pioneers often struggled to cope with the salty streams that flowed through their Eden. Freshwater salinity has accelerated in the past hundred years for all the same reasons dryland salinity has—an uncompromising farming system thrust upon a fragile land.

Professor Peter Cullen is Australia's foremost river ecologist. When asked how freshwater becomes saline he lifts two fingers up into the air like Winston Churchill: 'The salt is getting into the water through two main mechanisms,' he begins. 'It is coming from overland flow where we've got rain falling on salt-affected land and it's also coming into the rivers from groundwater, especially where the rivers are low. For instance, when the Darling is especially low, the groundwater starts to flow in and that puts salty water into the river.'

VIEWPOINT
Dr John Williams

JOHN WILLIAMS GREW UP ON A GRAZING PROPERTY IN NEW SOUTH WALES BEFORE STUDYING AGRICULTURAL SCIENCE AT THE UNIVERSITY OF SYDNEY, WHERE HE EARNED A DOCTORATE IN SOIL PHYSICS AND HYDROLOGY. HE JOINED CSIRO IN 1975 AND WAS BASED IN TOWNSVILLE FOR 16 YEARS. HE IS CURRENTLY DEPUTY CHIEF OF CSIRO LAND AND WATER DIVISION.

What we have done since European settlement is we've actually changed profoundly the water balance in the landscape. The original deep-rooted woody trees and shrubs and perennial grasses had come to a way of living with the salt. When we came along we tended to take them out and now a great deal more water dribbles past the root zones of our pastures and our annual crops.

Before we interfered with the system the amount of water that dribbled beneath the native root system was probably between half a millimetre and 5 millimetres a year. Now under our agricultural system of annual crops and pastures that number can be as high as 50, or in some cases even as high as 150 millimetres. So we have seen a tenfold change in the amount of water that is being pushed into the landscape.

If you push water into the landscape at a rate that is larger than the rate it can get it out, it must fill up and bring the salt into our rivers and into the lower parts of the landscape. Now when we decided to move to irrigation not only did we take the deep-rooted trees and shrubs out so we've got more water leaking below the root zone, we then also came along and put irrigating water on it.

No matter how skilfully we irrigate, it's very hard to stop water moving past the root zone of our irrigated crops, whether they are fruit trees or rice paddies or whatever. We end up putting more water past the root zone than the landscape has ever had before and more water than that landscape can move out quietly over thousands of years to the sea. So it's a very slow rate of release of water from below the root zone to the ocean and under irrigation we have increased it not tenfold but sometimes a hundred-fold, so we are the driving force for irrigated salinisation.

Perhaps the most important lesson in hydrology is that water is no pushover. Anyone who has ever picked up a baby's bath full of water and tried to carry it will attest to the weight of water and the pressure it creates. Imagine a hectare of land that has been cleared. For every 25 millimetres of rain that falls on that hectare, roughly 240 000 litres of water is absorbed. That weighs 240 tonnes. If that land is in a poor rainfall area and receives only 400 millimetres of rain in a year, it absorbs 3840 tonnes of water; if it is in wetter areas and receives 600 millimetres of rain then it absorbs 5760 tonnes of water. Year after year that water pressure builds beneath the surface. At the same time, as the trees have been felled and rain has been absorbed, the groundwater levels of the continent have built into a serious subterranean force. If for every action there is an equal and opposite reaction, then water pressure is the energy driving salinity.

In places that are flat and consistent—country that is plentiful in Australia—there is little drainage of the groundwater so these landscapes are susceptible to dryland salinity. However, if the landscape is varied then water will move downhill and sideways in the form of saline groundwater for hundreds of kilometres through the earth. One such body of water inches its way from the Grampians in western Victoria through to South Australia's Riverland region on the Murray River. Along the way it meets a large salt load stored in the earth and by the time the water seeps through cracks in the riverbank and into the Murray River it is like seawater. This is only one of countless underground waterways that are redistributing salt throughout the continent. Depending on how porous the soil is, water can move as much as several metres per day or as little as several centimetres per century.

Dryland salinity also plays a part in creating freshwater salinity. In landscapes where dryland salinity has left salt crystals on the surface, the vegetation dies away leaving the salt free to move either by wind or water. Few Australian waterways are still lined with native vegetation that once acted as a screen to filter incoming pollutants such as salt. So a heavy rainfall washes salt off the land straight into the water at the bottom of catchments.

Like so many others, Peter Cullen sees the inequity in the impact of freshwater salinity: 'Sometimes it's difficult to think about the impact of taking trees out. It may not appear on the same property, it might be 50 or 100 kilometres away. So with those distances it's very hard for humans to handle.'

When it comes to the scale of the Murray–Darling Basin, it is indeed hard to comprehend. It is a catchment area the size of France and Spain combined, stretching as far north as Charleville in outback Queensland, west to Broken Hill in New South Wales and south to the edges of Melbourne. The eastern border is formed by the Pacific Ocean. All up it covers one-seventh of the continent, an area of just over 1 million square kilometres (*see map, p37*) that contains more than 30 000 wetlands. All of the 24 major rivers and their tributaries in this massive catchment eventually drain into the Murray River. If it is difficult to grasp the scale of the catchment then it is all but impossible to comprehend the amount of salt under it.

Don Blackmore is the chief executive of the Murray–Darling Basin Commission, the body in charge of this vast and varied catchment. He is an engineer by trade and talks in a matter-of-fact way about salinity in the basin. There are four major groundwater systems in the Basin, the largest of which is the Murray Basin. According to Don Blackmore, there are 100 billion tonnes of salt lying dormant under the Murray Basin—a remnant of the old inland sea. 'It's a huge threat and I don't think it's reasonable to underestimate that threat. It's not terminal by any means—even if we didn't do anything for 20 years—but we will pay a price that we as Australians will find difficult to accept. So it's a huge threat, but I think as a society, we have had a call to action early enough to really make a difference.'

WHERE IS THE SALINITY?

Salt is everywhere in Australia but salinity isn't a problem everywhere—yet. Salt has been blown all around the country and transported in various directions by groundwater flow so its distribution is complex and irregular. The best estimates suggest

VIEWPOINT

Professor Peter Cullen

PETER CULLEN HAS BEEN INVOLVED IN FRESHWATER ECOLOGY
FOR MORE THAN 30 YEARS. HE ESTABLISHED AND WAS
CHIEF EXECUTIVE OF THE CO-OPERATIVE RESEARCH CENTRE
FOR FRESHWATER ECOLOGY BASED AT THE UNIVERSITY OF CANBERRA.
IN 2001 HE WAS NAMED PRIME MINISTER'S ENVIRONMENTALIST
OF THE YEAR.

Some parts of Australia are already unusable because of salt and soon more parts of it will be unusable unless we are able to do something serious about this problem.

I think it's interesting to reflect on how we made these mistakes. We sail in and make development decisions without really knowing what we are doing. We've cleared land and the government has encouraged people to do this. I think when that was happening, they didn't appreciate that taking out trees would bring the groundwater up and salinise the land.

It's a difficult area because there are big time-lags involved. It often takes 30 to 50 years after you remove the trees for the groundwater to come up and salinity to develop, and it's very hard for people to think of those time-scales.

We've known about the removal of trees and its impact on salinity since the early 1900s when the first papers were published, but people haven't wanted to acknowledge those things. We've been very keen to clear land and create wealth and our whole country has been about development and trying to live off our land. Well this is one of the real costs of that type of ignorance.

I think we have got some of the technical answers but we haven't got all of them. We certainly need to develop more comprehensive farming systems that will use up more of the rainwater.

One of the real paradoxes of this salinity problem in a dry country like Australia, is that by taking the trees out, we are virtually drowning the landscape. We have to find agricultural plants that will use the rainwater in the way that the native vegetation and trees used to—and wheat and pastures are not. There hasn't been a lot of serious work, in my opinion, in developing those new farming systems. There has been some effort but we need a much bigger effort in that area.

We need to get a better balance. We need to get a patchwork of vegetation—crops and trees and whatever— across the landscape that use the water that falls rather than letting it build up in groundwater. I am looking at catchment solutions where we address that hydrologic imbalance at the cause, rather than looking at the problem downstream.

there are about 2.5 million hectares of land that are already salt affected. A vast majority—70 per cent—of Australia's dryland salinity occurs in the southwest of Western Australia. South Australia is the third most affected state with dryland salinity widespread through the Eyre and Yorke peninsulas, Kangaroo Island, the Riverland and the southeast. Only about 1 per cent of Tasmania's farming country is affected by salt but across Bass Strait, Victoria ranks second only to the west with dryland salinity affecting the western Wimmera and Mallee districts and central Victoria. New South Wales has only 5 per cent of Australia's dryland salinity at present but is sitting on a time bomb with 50 per cent of the potentially affected land in the country and the largest number of people—urban and rural—who will be directly affected.

Salinity levels in creeks and rivers (including the Murray) vary seasonally. However, the Murray River collects an average of 156 tonnes of extra salt every day. Imagine 150 people backing a car trailer full of salt down to the riverbank every day and tipping it in. That happens 365 days a year. According to CSIRO's John Williams (*see also John Williams box, p58–59*), salinity levels in the Murrumbidgee River are increasing by up to 15 per cent per year and stream salinity in the Murray exceeds World Health Organisation levels for about 10 per cent of the year.

There is some remedial work that can be done to try to rehabilitate land that has been salt damaged but this is expensive and offers low returns. The smart science seems to be in finding out where the salt that will bother us in the future lies.

SALTY PREDICTIONS

In April 2001 a small aircraft lifted off from St George in western Queensland carrying equipment designed to search the earth below for signs of salt. Part of a multi-million dollar Commonwealth government-funded project (*see box, What is Airborne Geophysics?*), the search obtained information that was used to map potential saline hotspots in various parts of the country. Russell Speed from the Western Australian Department of Agriculture, says the hi-tech

WHAT IS AIRBORNE GEOPHYSICS?

In April 2001, the Commonwealth government-funded Airborne Geophysics Survey began in two selected catchments in Western Australia. This nationwide scan was marketed as the 'ultrasound of the earth' and formed part of the National Action Plan for Salinity and Water Quality.

Airborne geophysics for agriculture is a recent spin-off from the mineral exploration industry. It features three main components: magnetics, radiometrics and electromagnetics.

Magnetics measures variations in the earth's magnetic field and can provide a picture of the geology of an area; radiometrics, which measures the products of radioactive decay of naturally occurring minerals in the surface soil, can be used to help delineate soil types; and electromagnetics measures the electrical conductivity of layers of soil and weathered rock. Of the three, electromagnetics (EM) appears most valuable for obtaining information about salinity because of its potential to measure salt stored beneath the surface and locate underground streams, known as palaeochannels.

The technology involves a wire cable connected to the outside of an aircraft (aeroplane or helicopter), which 'pulses' an electric field into the ground below. This induces a secondary current in the ground and the strength of this second current— and the time it takes to decay—is monitored by a sensor behind the plane. Interpretation of these signals enables depth and conductivity of saline areas to be measured.

Because of the height of the aeroplane and a line spacing of 200 to 400 metres, the resolution of mapping is relatively low compared with that from ground-based EM systems, such as those carried on motorbikes—a survey unit familiar to many landholders. The data requires skilful interpretation in conjunction with other information such as soil and landform maps, groundwater monitoring and aerial photographs.

search was effectively a way of 'understanding catchment-hydrology processes'.

While scientists and developers agree it is important to know where salinity presently poses a risk, it is just as crucial to know where salinity is likely to spread. The simple way to do this is to find out where there are early warning signs of salinisation. In the years 1998 to 2000 the National Land and Water Resources Audit assessed the area of land considered to be at high risk of developing dryland salinity. It then used these figures to predict the area salinity might affect in 2050.

AREA OF LAND (HECTARES) AT HIGH RISK OF DEVELOPING DRYLAND SALINITY

	1998/2000	2050
New South Wales	181 000	1 300 000
Victoria	670 000	3 110 000
Queensland	not assessed	3 100 000
Western Australia	4 363 000	8 800 000
South Australia	390 000	600 000
Tasmania	54 000	90 000
Total		**17 000 000**

Source: The National Land and Water Resources Audit

Western Australia leads the charge with almost 4.5 million hectares affected and nearly double that expected in 2050. Richard George from the Western Australian Department of Agriculture says that amount will ultimately increase because he believes the salinisation process is only in its first trimester. 'I guess, generally speaking, we are about one-third of our way into the cycle that we hypothesise is taking place. About 1.8 million hectares of land is affected by salt. It's not all degraded because it's at various levels of salinity. About half of the water supply in Western Australia is affected by salinity and many species of native flora and fauna are at risk of extinction due to salinity. We have fairly large infrastructure threats. Rural towns have been developed in low-lying country because of the early settlement and establishment of railways and they have a fair risk of salinity. We are even suggesting that flood risks are changing as a result of larger and wetter areas in the landscape.'

Corey Watts from the Australian Conservation Foundation likes to use the analogy that, in the lifetime of today's school kids, an area more than twice the size of Tasmania is at high risk of salt damage: 'But the real problem is what lies in the wake of the salt,' he groans. Although his office is in the inner-Melbourne suburb of Carlton, Corey Watts frantically lists what he believes are salinity's environmental tragedies waiting to happen around the nation. 'More than 200 of our precious wetlands, many of them of international significance, are threatened. In Western Australia alone, scientists estimate that around 450 of that state's unique native-plant species are going to be extinct in the next few decades unless something is done. In South Australia, the Chowilla Wetlands stand to lose half the native vegetation. In Victoria literally hundreds of native plants and animals, many of them rare or threatened, are endangered by rising salinity destroying their habitat and in Queensland, where land clearing continues at breakneck speed, there is something like 3 million hectares of land that's going to be salt damaged.'

The scale of salinity can be difficult to grasp. The prehistoric salt harvest is measured in billions of tonnes and the hydrological changes that are driving the salt out of the ground and into our rivers, farms and cities are impossible to calculate. It is fair to say that every Australian faces a salinity issue in some form. Corey Watts calls that a crisis: 'Until we face up to this as a crisis we are not going to be able to minimise the full impacts of salinity on the Australian landscape and on Australian civilisation as a whole.'

So why does salt cause such a crisis?

CHAPTER 4
From Green to White

To the East or to the West,
Or wherever you may be,
You will find no place
Like the South Countrie.

For the skies are blue above,
And the grass is green below,
In the old Monaro country
At the melting of the snow.

BANJO PATERSON, 'AT THE MELTING OF THE SNOW'

There are so many breathtaking places in Australia that almost everyone can make a case that their corner of the continent is the finest. However, one of the most beautiful must be the Australian Alps. Here, towards the end of winter when the sky overhead is clear, the distant sun shafts through the eucalypt-forest canopy spotlighting the snow and illuminating crystals of ice that hang delicately off fronds. The forest smells clear and clean: tiny animal tracks are the only marks on perfect blankets of snow. Strips of bark hang like ribbons from the snowgums and twist in the breezes. Birds of all sizes and colours, from the brooding pied currawong to the hyperactive red-capped robins, call to each other. Past the birdcalls and the wind pushing through the leaves there is a small sound, so hard to hear at first that visitors have to hold their breath to listen keenly enough. It is the sound of the snow melting and the drops of water falling off tiny peninsulas of ice—the sound of the Alps giving birth to the Murray River.

It is a modest start for such a great waterway. The melting snow at the Pilot (1830 metres), a mountain in the Alps, is only the beginning. Throughout its journey 11 major tributaries and many more minor ones will bolster this body of water. The first join within a few kilometres of its beginning and the greatest is at Wentworth where the Darling River forms a junction that is the biggest freshwater event in the country. In between, it also collects water from the Lachlan, Mitta Mitta, Ovens, Goulburn, Campaspe, Loddon, Avoca and Murrumbidgee rivers. The Murray takes the long way home. It travels more than 2500 kilometres but the distance would be half that without all the twists and turns.

The freezing white water that first winds through Kosciuszko and Alpine national parks smoothes out and travels west through vast fertile landscapes, eventually forming into deep channels and then lakes. And then not 100 kilometres from its origin, the river stops as it reaches the first of what will be dozens of artificial barriers. The last of these is at Goolwa in South Australia.

A QUIET GOODBYE

If the birth of the Murray is a thing of beauty then its departure from this continent is ugly. Unlike the Mississippi that turns its deepest corner near New Orleans and bursts in one final surge through a delta into the Gulf of Mexico, the Murray limps through a thin channel into the Southern Ocean. A channel so small that Australian Conservation Foundation President Peter Garrett once gained attention by easily kicking a football across it to illustrate how pathetic it is. The gesture is hardly needed: anyone who looks at the waist-deep stream of murky water running through the sandhills knows that something serious must have happened to this waterway. Boats regularly become stranded on sandbars here as the water level drops, and long concrete and steel barriers fence the freshwater from the sea. These barrages are the last piece in an engineering system that has turned the river on its head. Contrary to nature, it now flows in summer rather than in winter because that is when irrigators need the water. Some argue it doesn't really flow then either because the rivers of the Murray–Darling Basin have been turned into a series of holding ponds. The only thing that passes as flow is when water is released from one pond into the next. The final barrier is Lake Alexandrina, where water can be held for up to a year before being able to trickle out into the Southern Ocean.

TURNING THE TAPS ON AND OFF

Lake Dartmouth is the largest storage dam in the Murray–Darling Basin and sits in the Alpine National Park in Victoria. It catches water from the park and in turn runs it about 50 kilometres down the Mitta Mitta River into Lake Hume near Albury in New South Wales. Although smaller than the dam at Dartmouth, Lake Hume holds six times as much water as Sydney Harbour and has a catchment that is more than 15 000 square kilometres. Built in 1936 Lake Hume was probably the most impressive engineering work on the Murray–Darling until the Snowy Mountain Scheme boldly went where no engineer had gone before. Lake Hume is such an imposing body of water that it has a yacht club and has hosted Olympic qualifying events. Here the water is still in marvellous condition: the reservoir gently fills from creeks and rivers snaking their way down the mountains. At this point the water being held back has travelled less than 5 per cent of its journey—but then much of it will never reach the Southern Ocean.

During winter, water collects and rises behind the dam's concrete and steel wall. Then in spring, the taps are opened and the water is let loose. Up to 25 000 megalitres (50 000 Olympic swimming pools) of water is released to supply 90 per cent of Australia's irrigation farming. Three million people from the Prime Minister in

VIEWPOINT

Dr Stuart Blanch

STUART BLANCH HAS A PhD IN FRESHWATER ECOLOGY FROM
ADELAIDE UNIVERSITY. HE WAS A CAMPAIGN OFFICER FOR THE
AUSTRALIAN CONSERVATION FOUNDATION BEFORE JOINING NEW
SOUTH WALES PARKS AND WILDLIFE SERVICE.

Before European invasion many species had to withstand periods of high salinity but the frequency and duration of those high salinity events are now increasing. So our fish and water plants don't get the breathing space they once got between periods of high salinity. That means it is getting more difficult to find times during low salinity periods to breed and to grow, and that's a major problem.

So it's disingenuous I believe to argue that salinity has always been around and so our nature can adapt to it. We have really changed the background levels. We are increasing the stress on our species and in many places they are not coping at all.

Too much salt in our rivers is just like another pollutant, such as oil or pesticides or too much agricultural runoff or fertilisers. As we increase that level of pollution in our rivers the species retreat more and more into smaller areas as a refuge from salinity.

A good example is the Macquarie Marshes, which are a jewel in the crown of the Murray–Darling Basin. The salinity level predicted to hit the marshes in 30 years is 1500 EC units, which is the level at which we are set to lose whole plant communities. Fish species are in real trouble and invertebrates have well and truly disappeared in many of the river systems. It's a real problem for us because to control the amount of salt entering that river system, we have to plant an enormous amount of trees in that catchment system including all along the western slopes of the Great Dividing Range, and we are not seeing that commitment by government yet.

We've got a lot to learn in Australia about managing some of our older irrigation areas. The average lifespan of irrigation areas throughout time and throughout the world is around 200 to 400 years. In the southern Murray–Darling we are about halfway down that path—80 to 100 years into that trajectory. We are certainly better at managing salinity now and better at farming. We have got more enlightened farmers for sure, and more money is going into farms for managing salinity, but we are also much better at ruining our landscape by controlling it and conforming it to our European identity, and that's a major problem. I think the jury is still out on how long our irrigation system can last.

Canberra to a rocket scientist at Woomera will drink from this water and with its help steel and plastic will be made and minerals dug from the outback.

The discharge will continue right through summer and autumn and the taps will not be turned off until about May, when up in the mountains the biting winds will again race through the snowgums and the river will prepare for its winter sleep.

Although the Murray River comes across its two largest dams within the first hundred kilometres of its journey, these are not the last. There are an estimated 3500 dams, weirs, locks, barrages, gates and sluices in the Murray–Darling system. The infrastructure might be starting to crumble in some places but it remains an engineering marvel. At Blanchetown, about an hour's drive from Adelaide, Lock One was built in 1922. It remains almost intact despite being opened and closed several times a day for more than 80 years. The locks and weirs, which were completed by 1930, helped smooth out the river's flow. No more boom and bust to upset riverboat captains and irrigators could be guaranteed delivery of their water. The Murray River may start naturally but pretty soon it is distorted almost beyond recognition. It now flows when it shouldn't and is quiet when it should be noisy.

'On top of the unruly, unpredictable, variable Australian river system, we as European and Middle Eastern descendants brought our agriculture and our way of thinking and largely plonked it on top of our Australian river system,' argues environmentalist Stuart Blanch. As the son of a New South Wales banana grower, Stuart understands the farmer's perspective and respects the engineering often involved, but armed with a doctorate in river ecology he is fighting for the rivers to be set free. 'We have tried to make nature conform to the way we think rivers should operate, and the way the crops we bring in should operate, and it hasn't worked to a large degree. For example, in the Murray–Darling Basin we have 30 large dams. We have enough storage in that basin to capture about two and a half times the average annual runoff into the river system. So we are very good from an engineering point of view at controlling the river systems and using them as a supply canal to deliver water to the crops, which grow largely in summer. Now that's a problem for the fish and the plants and the birds and the wetlands because we have changed the rivers.'

SALT COMES ALONG FOR THE RIDE

In 2000 the remarkable marathon swimmer Tammy Van Wisse became the first woman to swim the length of the Murray River. Part of her plan was to draw attention to the state of the river and through daily media interviews she gave a first-hand account of its health. She remarked how invigorating the freezing headwaters were to swim in, and how the waterway became lethargic and filthy as she continued. She needed support staff to paddle alongside her in a kayak to fend off logs and snakes because she couldn't see her own hands in front of her. By the end of her triumphant swim, Tammy Van Wisse says she could taste the salt in the water. It wasn't from the ocean mouth because the barrages keep the salt and fresh water separate. It was because Australia's rivers, like its land, are getting saltier.

Tammy Van Wisse had the advantage of being able to get out of the water after her two daily swims, have a shower and drink bottled fresh water. It isn't so easy for the natural river users. The swimmer made sure she always covered herself in sunburn cream before setting out. The well-drilled message with skin care has been that by the time you start to feel it, the damage has already been done—so too, with salinity. Humans start to cough and splutter if they taste water with a salt level around 800 EC or 500 milligrams per litre (*see box, How are salinity levels measured in water? p76–77*). But for many of the wetland plants the problems are well advanced by that stage and in some cases irreversible damage has already been done.

Wetlands are often referred to as the 'kidneys' of waterways because they clean and purify water. Bacteria and other microbes breed in wetlands and while they can clean most things, they cannot eliminate salt. Many wetland plants start to have lower breeding success and seed production when salinity levels exceed approximately 600 EC. This level of salinity is commonly reached

HOW ARE SALINITY LEVELS MEASURED IN WATER?

Measuring salt levels in water is easy but the way the results are recorded and interpreted can be confusing. There are three basic units to record water salinity but the use of these units varies from state to state and sometimes even between agencies.

The concentration of salt is basically the weight of the salt in the water. This is measured in parts per million (ppm).

Fresh water = less than 1000 ppm
Slightly saline water = 1000–3000 ppm
Moderately saline water = 3000–10 000 ppm
Highly saline water = 10 000–35 000 ppm

Seawater is at the very top of the scale at 35 000 ppm which means about 3.5 per cent of the water is salt.

Another way of expressing salt concentration is milligrams per litre (mg/l). Roughly one part per million is equivalent to one milligram per million milligrams (1 kilogram or 1 litre).

Salinity is measured in water by testing the electrical conductivity. It works on the simple principle that salt water conducts an electrical current far better than fresh water. A cell is dropped into the water and an electrical impulse passes through it at a certain voltage. If the temperature of the water is known and the pH (the acidity of basicity) is in the normal range (close to neutral) then the conductivity measurement can be used to estimate the salinity within about a 5 per cent error.

This measurement of electrical conductivity (EC) is expressed in micro Siemens/cm. Murray–Darling River irrigators have grown used to using this system of measurement and can apply the readings to estimate how the water will affect their crops and soils. The relationship between salinity and conductivity is not a linear one, but at relatively low levels such as in drinking water it basically follows this formula:

$$EC \text{ (micro Siemens/cm)} \times 0.548 = \text{total salinity (mg/l)}$$

Salinity levels in Australian drinking water vary but the average is around 50 mg/l and most people can begin to taste salt in water when concentrations reach 180 mg/l. The Australian Drinking Water Guidelines suggest the following guide to drinking water:

0–80 mg/l = excellent
80–500 mg/l = good
500–800 mg/l = fair
800–1000 mg/l = poor
1000+ mg/l = unacceptable

Professor Don Bursill from the Australian Freshwater Co-operative Research Centre says the guidelines are set around aesthetics not health requirements: 'So you have to go very high before there are any adverse health affects.' The World Health Organisation sets down 1000 mg/l as acceptable for drinking. Murray River water in South Australia is currently 350 mg/l while rainwater is 80 mg/l.

Salinity can also be measured by analysing water samples for all of the constituent inorganic components (sodium, calcium, chloride, sulphate etc.) This can be done in laboratories but is more expensive and time consuming.

Source: Freshwater Co-operative Research Centre and South Australian Department of Water Resources.

and exceeded in Australian waterways. These levels of salinity may not kill the plants outright—that occurs at about 8000 EC—but over time and successive generations they will decline in number and vigour. A wetland may have a salinity level of 500 EC in late spring when water is flowing, but by the end of summer when the water reduces and evaporation has taken its toll, the salt levels may be much higher. Micro-organisms suffer, frogs suffer, and ultimately birds and fish suffer from lost habitat. Although river fish have evolved from marine ancestors and can handle changes in salt loads, they can't thrive if the load just increases. As South Australian research scientist Bryan Pierce says, 'If you want fish then add water.' River flows have been reduced so much in the Murray that commercial fishermen had to stop using drum nets. Such nets are like long wicker barrels with a funnel at each end. They work on the theory that if enough water flows through, then fish will be caught inside the net. When they are full they are raised into a boat from where the small fish are tossed back. Commercial fishermen have used drum nets since the late nineteenth century, but in the early twenty-first century they became museum items because changes to the river flow meant they couldn't catch anything.

CHANGES IN RIVER FLOWS IN THE MURRAY–DARLING BASIN

River basin	Flows under natural conditions, in GL/Year[1]		Current flows under regulated conditions in GL/Year[1]	
	Mean	Median	Mean	Median
Darling	3042	1746	2272	1053
Murrumbidgee	2794	2527	1184	644
Goulburn, Broken and Campaspe	3668	3510	1774	1211
Loddon	247	202	100	37
Namoi	872	570	402	177
Gwydir	60	11	120	55
Murray	13 754	11 883	4915	2539

1. GL = gigalitres
Source: Murray–Darling Basin Commission

Stuart Blanch explains that our native species have adapted to a boom–bust river cycle that included the highs and lows of salinity first recorded by Charles Sturt. However he doesn't like their chances of surviving the new system. 'In some periods we would have floods two and three years in a row, then maybe no floods for five or six years, and these species would cue or respond to the big floods. In the Murray–Darling Basin where we have 26 species of native fish, a lot of those such as the Murray cod, yellow belly and silver perch, would respond to big floods. They would have enormous breeding events so you would have very good reproduction in those periods but then perhaps not for another five or six or ten years until another big flood came along. So they needed the rise in water level to breed. In addition, the invertebrates that fish rely upon to survive were affected by changes to flow.

'The changes we have brought on the river system now through regulation and farming have to be viewed in the light of that natural function. And that's a problem because nature hasn't adapted to the way that we now manage our river system. The symptoms of those changes have been astronomic to say the least. In the Murray–Darling Basin where we've lost something like 50 per cent of our wetlands, we are losing a lot of our fish and our waterbirds are retreating just because we cannot provide them with the water they need.'

The waters most affected by salinity in eastern Australia are in the southern part of the Murray–Darling Basin. Although Western Australia doesn't have nearly the amount of water as the east of the country, salinity is also a significant problem. A study in 1996 by CSIRO into the state's southwest showed 80 per cent of the length of rivers and streams were degraded by salinity and half the waterbird species had disappeared from wetlands.

But it isn't only the rivers and wetlands that are suffering.

THE PLANT DROUGHT

If young Australians want to see what their country looked like 200 years ago they often have to visit national parks. Like museums, they give a glimpse into a world that once was. So many native

VIEWPOINT
Rachel Siewert

RACHEL SIEWERT HAS A BSC IN AGRICULTURE AND HAS LIVED AND WORKED IN RURAL WESTERN AUSTRALIA. SHE HAS BEEN CO-ORDINATOR OF THE CONSERVATION COUNCIL OF WESTERN AUSTRALIA FOR MORE THAN 15 YEARS.

I resist very strongly putting a dollar value on the biodiversity that we have lost. Its value for me is priceless and incalculable. You can put a dollar value on some things like infrastructure and agricultural production that will be lost to salinity, but you can't with biodiversity. It's got a right to exist as far as I am concerned and this planet will be a much poorer place if we lose these things.

What we really do need in Australia is a more comprehensive approach to natural resource management. We also need to recognise that the level of investment we have had in the repair of the country is infinitesimal compared to the amount that we need to invest. Unless we get serious about the level of investment we are wasting our time to a certain extent.

We need much stronger leadership in getting that investment and obviously not all of that can come from government, we appreciate that, but unless we get the leadership from government I don't think we are going to see the investment coming from industry or individuals.

You have to look at how we prioritise our investment. What are our major assets in infrastructure, farmland and biodiversity? We need to ask should we be investing our money in building up new options that deal with the problem, or do we invest them in the immediate crisis? That's the thing we are coming to terms with in Western Australia. I don't think, at this stage, we have a very clear way of determining where we should be investing our money in protecting our landscape and biodiversity and our agricultural land from salinity. We haven't quite nailed it yet but at least we recognise we need to do it and it's a start, and I think some other states haven't done that yet.

We have to change the way we manage the landscape, we have to change the way we do agriculture. We have to find new crops to replace the shallow-rooted annual crops with deep-rooted perennial vegetation, and have agricultural systems that make money out of those systems. In other words we need to address this issue on a commercial basis. We need incentives that help farmers make a buck and that can start affecting salinity because it's pie-in-the-sky stuff if we think farmers will do this for altruistic reasons—they just can't afford to, it's plain economics.

In Western Australia we are maturing in the way we are dealing with it. I think we are seeing an interesting phenomenon, and that is a lot of leadership is coming from farmers who are interested in natural resource management. They are the ones who are starting to lobby government and they are the ones who are starting to lead the push for this new agriculture. I don't think we are seeing that at a national level yet.

landscapes have gone and with them the biodiversity they carried—the birds, reptiles and animals. Remnant native vegetation is also under threat from salinity. Western Australia again is the state with the most distressing forecasts.

Rachel Siewert has been involved in studying and lobbying for the Western Australian environment for almost 20 years. On a sunny winter morning she slips out of a conference seminar at the Department of Conservation and Land Management in Perth. Across the road is a strip of natural bush and as she settles on a fallen log to discuss her concerns, only the occasional jet overhead reminds her she is in a major city. She begins with an explanation of what the state had looked like before the axe and diesel got to work. 'In the northern plains we had a lot of heathland vegetation that contained extremely high levels of biodiversity. We had low level woodland in the central wheatbelt, hardly any of which is left. It must have been absolutely stunning. Then in the south coast you had a lot of mallee-type vegetation, which again was highly diverse. The southwest of Western Australia is one of the most biodiverse areas in the world and unfortunately we've already lost so much of that. About 22 per cent of the plant species of the southwest of Western Australia are extinct or threatened so they are under a tremendous amount of pressure.

'Salinity is having a very significant impact on biodiversity but what we are most concerned about is the future predictions for the biodiversity we will lose. For example, 450 endemic plant species that have been identified in the wheatbelt, will be lost. There are further regionally significant species beyond that which are potentially going to be lost. There is also fauna associated with those plant species. There are a lot of important freshwater lakes still remaining and of the 61 waterbird species that use them, only 16 can tolerate hyper-saline conditions. There are a number of invertebrates that are often forgotten that will be devastated as well. So basically the picture for biodiversity survival in the southwest is very bleak.' (*See also box, Rachel Siewert p80–81.*)

The bleakness is because salt is so destructive to plants and without plants there are no ecosystems. 'What we are doing essentially is making the plant suffer a drought,' explains John Williams. 'The salt has an osmotic effect and by osmotic I mean it dries out the water in terms of the plant's ability to extract it. Also the excess amount of sodium causes an imbalance in the other nutrients the plant needs such that the plant suffers nutrient deficiency as well.'

Small doses of salt won't kill a plant but they do weaken the plant's defences and make it harder for the plant to recover from grazing, diseases, drought and poor soil nutrients. Plants with leaves that are in contact with water are more susceptible as the salt in the water is constantly absorbed by the plant tissues. Due to osmosis (water molecules moving from areas of low salt concentrations to high concentrations) the salt in the water can virtually suck the fresh water out of the plant. Plants are most susceptible to salinity when they are seedlings because they have small roots and the leaves are often fleshy and easily damaged by salt.

The rate of salinisation in Australia was increased astronomically by the destruction of native vegetation. The second act of this tragedy is that the native vegetation that is left will continue paying the price for the salt that's been unleashed.

CHAPTER 5

When Salt Comes to Town

Every great mistake has a halfway moment, a split second when it can be recalled and perhaps remedied.

PEARL S. BUCK

Wagga Wagga is the largest inland city in New South Wales and sits on the banks of the Murrumbidgee River almost 500 kilometres south of Sydney. It's beautiful to walk along the riverside in any season. In winter the water is slow and dark but in summer it surges along with the sun dancing off it. The city boasts excellent botanic gardens, a university, art galleries and sporting grounds. Further out are prosperous farms and two military bases. Increasingly however, people come to visit Wagga Wagga for another reason. Tourists of a different kind wander into the gleaming Civic Centre and ask for a booklet with all the information and maps needed for a self-guided tour of urban salinity.

Bryan Short from Wagga Wagga City Council has entertained countless people in his first-floor office who all say the same thing—tell me about urban salinity. 'When we look back on what happened there were probably signs there 20 or 30 years ago but we really only recognised it for what it was back in 1993. There was redevelopment work taking place at the local showgrounds where they were constructing a new trotting track and that involved a cut-and-fill earthworks operation. After the trotting track was finished the show society tried to re-establish grass on the arena and had difficulty doing that. They sowed the grass once or twice and it died both times. We got a backhoe up there and dug a few holes in the arena. In a very short space of time the water had built up to within 10 centimetres of the surface and when we tested the quality of that water it was about one-third of the level of seawater. So it was then that we really realised that we did have a problem of shallow groundwater across a large part of our city.'

Bryan produces a series of photos that show what happens when salt comes to town. There are images of parks that are littered with bald spots from salt scalding, crumbling bricks on buildings and road gutters that have a trail of salt in them. 'The impact on the city was that it was starting to cause damage to brick works. It was reducing the life expectancy of our roads from about 30 years to about 10 to 15 years. We were starting to see our sporting fields being affected because the vegetation was dying off. We've had a number of economic studies done, the most recent one indicating that if we didn't do anything to ameliorate the problem the community would incur costs of $180 million over the next 30 years.'

Wagga Wagga is only one of 68 towns in the Murray–Darling Basin that has an unhealthy diet of salt. It isn't a fashionable thing to admit you have sodium chloride as a neighbour so in many places heads have been buried in the sand. Not so in Wagga Wagga where the city has been upfront about the problem, tackling it head on to give themselves the best chance of managing it. Given the expectation that within 50 years 200 towns in the basin will be affected by salt, you can see why they have more than a few civic neighbours picking up an urban salinity tour brochure.

WHY WAGGA WAGGA?

Wagga Wagga is built on a floodplain of heavy clay that was once covered with woodlands of river red gums, cypress pines and wattles. There are ancient salt deposits lying in the soil and poor discharge for groundwater. In other words, all the ingredients for a classic case of dryland salinity should the system be upended—as it was. Farmers began the process by clearing land and allowing watertables to rise. Then a city was built on the floodplain and, as the city grew, it spread from the floodplain and up into the surrounding hills that channel water down into the catchment. Deforestation had robbed the land of its natural pumps, and rain that fell into the higher country soaked into the ground and forced its way down into the watertables of the lower areas. Soon the heavy soil became waterlogged and the land couldn't drain itself.

Many houses also had poor drainage systems. Rather than running stormwater away from houses, the water was effectively stored on site in rubble pits. These were large pits dug in the backyard and filled with gravel or broken bricks. Stormwater was then channelled out of roof gutters into the pits where it drained away. The trouble was that, instead, it added to the watertable that was building like a giant water-filled balloon under the city. As the watertable rose, it absorbed the ground salt until eventually this salty build-up found a way out. That is why when they picked at the surface of the showground's trotting track, workers found seawater.

Wagga Wagga is a proud sporting city. At one point in the 1990s it could boast that it was the birthplace of Australia's cricket captain and his opening partner, Mark Taylor and Michael Slater; the captain of the AFL Premiers North Melbourne, Wayne Carey; and Brownlow medallist Paul Kelly. The city lies close enough to the Victorian border that champions have been produced in all football codes. Near the outskirts of the city though, there is a sad sporting sight. A rugby field that no one can play on because salt has killed the vegetation. Even during dry weather there are salty puddles and the surface is sloppy.

VIEWPOINT
Bryan Short

BRYAN SHORT IS AN ENGINEER AND DIRECTOR OF
ASSET MANAGEMENT AT WAGGA WAGGA CITY COUNCIL.
HE HAS BEEN A PIONEER IN IDENTIFYING URBAN SALINITY AS AN
ISSUE AND DEVELOPING PROGRAMS TO ADDRESS IT.

The sort of tell-tale signs of salinity in Wagga Wagga were dampness under houses, vegetation dying, roads breaking up, salt in table-drains and in gutters. We went through a three-year assessment phase and at the end of it we realised that we could tackle it by using four programs. The first was education, the second was revegetation, the third was what we call a leakage-reduction program where we try to reduce the urban-type inputs into the watertable by leaking water supply systems and through rubble pits, and the fourth program was pumping, where we put nine bores through the area most severely impacted from the rising watertable.

We think the dollars we are spending on education are probably getting us the best results because it is not a high-cost program but it does have a substantial benefit. In terms of revegetation we set ourselves a target of revegetating 20 to 25 hectares per annum and to date we have been able to exceed that. In leakage we have a pilot area where blocks have new drainage and their internal supply is checked for leaks. The bore field project is seeing positive results with the watertable lowered by 2 to 3 metres in some areas.

We have a booklet recommending what people should do when building in a saline environment. To date those recommendations are only advisory. We've eliminated rubble pits and if people can't get stormwater out to the street and there is no rear-block drainage then they have to put storage wells in their backyard and pump it out into the street.

Our rising watertable is not a regional problem in that the problems are not in the floor of the Murrumbidgee Valley but up in the sides of the valley where the clay soils are. I think the main reason for salinity in the urban area is the urban activities that took place in the early part of the century. Along with widespread clearing there was the introduction of roof drainage systems that involved rubble pits and water supply systems where there wasn't a lot of attention paid to keeping leakage to a minimum. If a community can address all those issues in urban areas then I think they can keep salinity under control.

'Classic dryland salinity,' says Greg Bugden from the New South Wales Department of Land and Water Conservation. Greg doesn't need the city's brochure to know where the salt lies; he has lived in the area long enough to know the hot spots. His tour begins at the edge of the rugby field where he points at cracks in the road. They are weeping like sores and as the sun hits the moisture it evaporates leaving a small ring of salt behind. Nearby roads have been patched with extra tar and in some areas the number of patches applied over and over to stop the salinity is almost comical—bitumen bandaids. 'What we are seeing is the road breaking up. Underneath the road the clay is saturated and the moisture and salt moves the tar. The tar loses elasticity and starts to crack up. [A] road could cost in excess of half-a-million dollars per kilometre to fix. If it's a highway then it will probably be a million dollars per kilometre. Rather than lasting 15 years it only lasts six.'

A short drive away is an abandoned red-brick building that Greg uses as an example of how salinity can eventually destroy the strongest domestic structures. 'Look at the high tide mark,' he says as he squelches across the salty wet area that used to be a front lawn. Sure enough, more than a metre up the wall, is the line where the water has reached. 'You see bricks are like sponges,' he says tracing the bricks from the ground to above his head with his finger, 'and they pull the moisture right out of the ground.' Greg picks at the face of a brick and a chip the size of a credit card falls away. Underneath is a mixture of red brick and white salt. He steps back from the brick face and squats down on his haunches to finish his lesson in urban salinity: 'At this location the watertable would be about half a metre below the surface. With the capillary action the moisture and salt are brought to the surface and the vegetation dies off. The result then is that the bricks, being very porous, are able to draw the moisture and salt up. After the second layer of bricks you have a dampcourse but it's breaking down, allowing the moisture to move up the wall. The result is salt crystals building upon salt crystals. Eventually the bricks become very soft and defoliate. Over a period of time, maybe five to ten years, the whole brick will break down and you'll be able to put your fist though it. So the cost would be in the order of $10 000 to $15 000 to fix this problem. What we need to do is replace the dampcourse and maybe render over this area, but most importantly, try to lower the watertable to below 2 metres so that the process doesn't carry on.'

While homeowners try to reverse the decaying effects of urban salinity, the city has aggressively gone after the fundamental cause. Rubble pits have been banned and new drainage installed in areas that are badly affected. In high recharge (*See Water Cycle page 12–13*) areas hundreds of native trees have been planted to try to reduce the amount of water flowing underground to low-lying areas. Initially there were security concerns from residents about the idea of creating a forest in the middle of their street so barbecues and public meetings were held to teach them of the need for such vegetation. Wagga Wagga's outgoing approach towards public education means right across the city there are signs explaining what is going on and why. There wouldn't be a citizen who is unaware of the problem or the solutions and perhaps there is even a sense of pride in the city's accomplishments. The most obvious effort though is the nine production bores in the centre of the city that keep the watertable in check. At the moment the salt concentration in the groundwater is low enough that the excess can be pumped into the Murrumbidgee River. These bores have combined to lower the watertable to 3 metres below the surface. At that level, the wall that Greg was able to pick apart like a pavlova shell will eventually dry out. However in such an extreme case he believes the damage is terminal: 'The salt would have to be flushed out and the bricks replaced but I think the building has gone too far for that to happen. What we should be doing with new buildings built in saline environments is to incorporate exposure-grade bricks below the dampcourse and make sure the dampcourse is satisfactory.'

PREVENTION IS BETTER THAN CURE

On a windswept block a group of builders are putting together the latest addition to a new Wagga Wagga subdivision. This is where Greg Bugden believes the final piece of the anti-salinity battle is being fought. A thick plastic membrane is laid down before the concrete

WHAT CAN YOU DO TO STOP URBAN SALINITY?

- Reduce the amount of lawn around your home.
- Plant native trees and shrubs on vacant land.
- Water gardens only when necessary. A sprinkler can use 1000 litres of water per hour so turn off automatic sprinklers over winter.
- Use a timer with sprinklers or a micro-irrigation system.
- Use raised garden-beds and import fresh soil.
- Reduce evaporation by adding mulch to garden beds.
- Group plants with similar water usage together.
- Lime and fertilise only as needed to make the most of water used.
- Select salt-tolerant species.
- Test your soil to find out the salt load.
- Use potted plants if the soil is too salty.
- Fix any leaks in water pipes and check your swimming pool for leakage.
- Let council know of parks and gardens that are overwatered or where leaking pipes are suspected.

Source: Wagga Wagga City Council and NSW Government Salinity Management Program.

foundation slab for the house is poured. This is a simple way of stopping the damp rising. Later another strip of plastic is rolled out on top of the first layer of bricks as a second protective layer. Bricks and pavers that have a higher resistance to water are also being used. These protective measures are all optional at the moment but Greg Bugden is among many urban planners who would like to see them made a mandatory part of the Australian building code. 'The costs we've looked at here in Wagga should be only about $4000 per home, so spending that $4000 today may reduce the cost of $10 000 to $15 000 per home in 10 to 15 years' time when salinisation occurs. All up in Wagga Wagga, salinisation could cost $183 million over the next 30 years. Some of the big items are roads, which will cost $56 million, and repairs to homes which will cost around $20 million.'

A visit to the showgrounds today reveals a lush, green lawn inside the trotting track. Nearby signs explain how subsurface drainage lowered the watertable by draining the saline water to a nearby evaporation pond. The pond is 2.5 hectares in size and 1.5 metres deep. The exercise wasn't cheap—all up it cost $300 000 to craft the solution. That appears to be a very small amount when compared to the urban salinity bill facing the nation because it isn't only a rural problem. Salt is creeping into the biggest cities of the nation.

THE BIG END OF TOWN

It might be fair to think that the urban salinity problems being faced in rural and regional Australia are the result of land clearing for farming, and this is true to an extent. In addition to those towns and cities in trouble in the Murray–Darling Basin, there are over 30 towns in Western Australia currently trying to manage salt infestations and another 30 expected to be affected in the future. However if cutting down trees, burning scrub and farming the land is considered a radical change, then it is nothing compared to building a city. In a city not only are trees cut down and bush burned, but millions of homes, factories, office blocks and highways are built on top of the land. Each one acts like a minicatchment with every hard surface from roofs to pavers providing a fast exit for the water. Good urban planning involves building a system to get rid of this water, and sewers, gutters and drains do just that. Most ends up going into the ocean very much as nature might have planned it. However nature didn't intend every owner of the red-brick dream to plant a 'farm' out the front and back of the castle. Australians love lawn with a passion. When they built the first federal Parliament House in Canberra in the 1920s it seemed to reflect the theme of the day: Ben Chifley could look out of the prime ministerial office window at a mob of sheep grazing nearby. The new building, by contrast, features a paved courtyard to represent the outback, columns of grey marble for the gum trees, a tapestry the size of a tennis court showing virgin bush in New South Wales and a shiny steel and neon coat-of-arms and flagpole that seems to indicate a bright future on the cutting edge—but the building itself is covered with a lawn that rivals the MCG. There is a brilliant unintended explanation of Australia in all that.

'Of course it's not that there is anything wrong with lawn—except it can't hold its drink,' says Rebecca Nicolson, the salinity project officer with the Western Sydney Regional Organisation of Councils (WSROC) that represents one of the fastest-growing urban areas of Australia. 'What we tend to do in urban development is use a lot more water, we water our gardens and our lawns, we build concrete structures and the roofs have a lot more runoff. We also have pipes under our houses that sometimes leak and that adds more water to the system.'

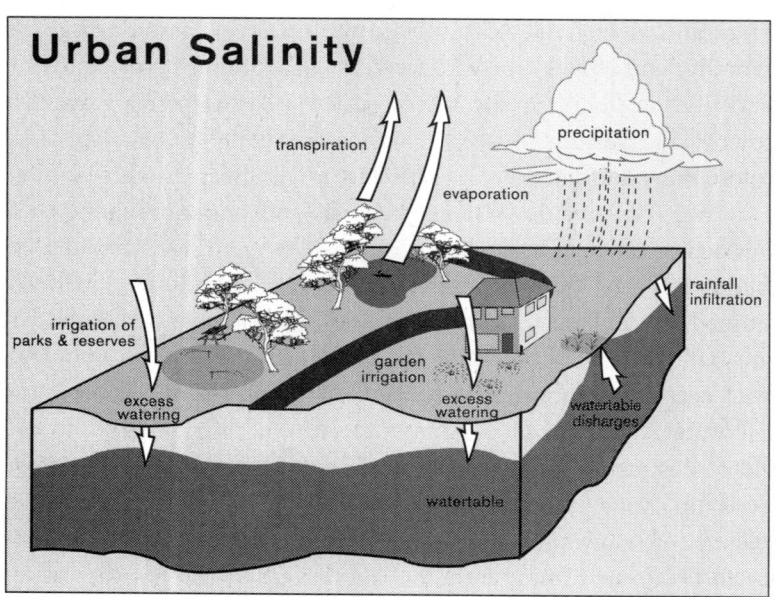

Urban Salinity

In urban areas clearing of native trees along with excess watering of parks and gardens, leads to a rising watertable bringing salt with it. Bricks and mortar absorb the salt which ultimately eats away at them.

If western Sydney were a city on its own it would be the third-biggest in the nation and its economy would rival Singapore's. It has a population of more than 1.5 million people and every day new houses are being built and the city limits pushed further and further into the bush. If people who live in the west feel a little disconnected from the more established east of the city, then so does the soil. While Sydney's eastern soils evolved from coastal sandstone, further in the west the soil gets heavier and more clay-like. This means there is more salt in the soil and the potential for urban salinity is much greater. If that salt is mobilised then there are more than a quarter of a million homeowners praying that their dampcourse works.

Already there are some signs of salinisation. A drive with Rebecca reveals the odd carpark with a bubbling blacktop effervescing salt, bricks crumbling in some older buildings and parks with salt scalds. She says the catchcry for western Sydney is prevention. 'We have looked at what is happening in places like Wagga Wagga and we have made a decision that we don't want to go down that track. So our councils are committed to developing a proactive response—to manage the problem now while it is just a ground problem.'

Rebecca rolls out a map of the region and points out what she refers to the 'salinity hot spots' that have been identified: 'What the hazard maps are telling us is areas where there is an actual salinity outbreak and the distribution of them. They are also telling us areas where there is a high potential for a salinity problem. They indicate that basically all of western Sydney is at risk from localised salinity hazards and that is because the soils are saline. If you create damp situations through leaking pipes or structures that impede soil drainage, you can create a salinity problem.'

Greg Bugden has looked at western Sydney and has an equally blunt response when asked about its future. 'The community of western Sydney should be asking themselves: Where is the watertable under my home and its vicinity? What can we do to keep it at that level rather than leaving it for another 10 or 15 years and relying on engineering solutions which are very expensive?' He applauds the hazard mapping and believes the western Sydney councils are in a great position to protect sensitive areas from urbanisation.

Together the 13 councils are trying to manage the salinity hazard by consulting developers and homeowners. Camden is one of the fastest-growing areas on the outer fringe of western Sydney. To the west the view is rural but to the east there is suburbia for more than 80 kilometres until it hits the Pacific Ocean. The council is involved in a delicate dance trying to tiptoe around the salinity hazards as they waltz with development. They have held forums and public meetings and invited builders, consultants, council workers and residents to try to broaden everyone's knowledge. Urban planning has included trying to avoid cutting across natural contour lines in the land to reduce the amount of cut and fill needed. Corridors of significant vegetation are left intact and that means leaving small forests rather than single trees. Homeowners are being asked to install rainwater tanks to reduce the amount of water going into the creeks and then infiltrating the ground. One developer has agreed to provide a native garden in the front of every house he builds in the area, and council wants the neighbours to follow the trend in both their front and back yards. They've even asked builders to consider moving away from slab-on-ground construction and return to post-and-beam or pier-grid construction so the house foundations are moved off the ground altogether. Cathy Kinsey from Camden Council agrees with Wagga Wagga's assessment that education is the best weapon and she uses the great mantra of Australia in the twenty-first century—sustainability: 'We are looking at a total catchment management approach. We are looking at coming up with a sustainable solution that will provide us with some protection from salinity both now and into the future.'

Sydney's weather goes in its favour when it comes to salinity. Its winters tend to be dry and the hot months of the year are wet and humid. This means the evaporation rates are relatively low during summer and the salinisation process is slower. There is one city that has the reverse situation. In Adelaide the winters are cold and wet and the summers are dry and hot. The winds that bake the city in summer come from the desert to the north. Evaporation rates are very high. The city is built on a flood plain nestled between a mountain range and a gulf so the soil has plenty of salt in it. If Sydney residents want to see how to treat urban salinity then they should take a trip south.

CITY OF SALT DAMP

'Salt damp' is a phrase that every Adelaide homeowner is aware of but one that is almost unknown outside South Australia. In other states the phenomenon is known as 'rising damp'. It means the same thing: groundwater and the salt it contains is being absorbed by bricks and causing damage.

The current site of Adelaide is only one of several locations scouted for a colonial capital in 1836. There was a push for the site on which Port Lincoln on Eyre Peninsula now stands because it had a better harbour (Boston Bay is three times larger than Sydney Harbour) but even at that early stage, water was an issue. Port Lincoln and the other possibility, Encounter Bay on the Fleurieu Peninsula, weren't considered to have a reliable enough water supply. So the flat country occupied by the Kaurna people was chosen.

Almost as soon as the site was decided upon, the first boatloads of migrants arrived from Britain. There was enormous pressure on the Colonial government surveyor to get on with the city design so that land promised to settlers before they left England could be handed over. Colonel William Light was no stranger to pressure though. His father was also a city planner, having designed Georgetown, the capital of the Malaysian island of Penang. He lived with his father and his Asian mother only until he was of school age, after which he was packed aboard a ship and sent to England. He later served in both the army and navy and fought in the Peninsula War and in Egypt. One of his most enduring characteristics was a single-mindedness—even to the point of ignoring public scandal by living openly with his mistress Maria Gandy. As William Light slashed through the bush with his team surveying the new city he didn't know the amount of salt that lay in the soil beneath his boots. He was busy fighting tuberculosis and raging against his arch-rival the South Australian governor, John Hindmarsh. Within a few years the city was pegged out and Light was dead at 55. His legacy was a sympathetic and efficient city laid out around parklands and the Torrens River—but one that quickly began to reveal its salty character.

The soil in Adelaide contains a lot of salt but coupled with the added problem of high evaporation rates during the furnace of summer, South Australians have had to deal with urban salinity for a long time. Tackling salt damp has become a way of life, beginning with drainage and ventilation and ending with dampcourses. Water should always be drained away from the house so that it doesn't pool near stonework or bricks that will absorb it. Some people lay concrete around their homes or even plastic sheets beneath the surface so that no moisture seeps into the ground but instead runs away from the house. Gutters, drains and water pipes are maintained so they don't leak. Garden beds are not allowed to build up next to the house and certainly never above the dampcourse. In the early days, dampcourses were made of materials like ceramic tiles, slate, glass, lead, bitumen or tar. A row of this material was laid on top of the first row of bricks so that any moisture being drawn up would stop at this low level and evaporate out without weakening the structure above. Now dampcourses are made of polyethylene.

However, as the people of Adelaide know, even the best precautions can still fail. To the horror of the homeowner bricks will flake, plaster walls bubble and walls collapse as salt grinds its destructive path through the house. Several techniques have been developed to overcome the damage beginning with a new dampcourse. The most radical surgery is to underset sections that have been damaged. This involves cutting the old bricks and foundations out section by section and replacing them with new salt-resistant materials. A gentler almost keyhole-type operation can be just as effective. First holes are drilled into the walls and then silicon-type mixtures are injected into the walls. They spread and form a waterproof layer within the wall. If there is moisture and salt above this layer it will evaporate out but no more water is able to rise. However if the salt that has already risen above the new dampcourse has caused a large amount of damage, then there are remedies available. So-called 'sacrificial renders' can be applied to the walls. They are made of highly porous materials that are attractive to moisture and salt. Once the render has drawn the saline water out of the wall, it can be chipped off like a plaster, leaving the dry wall behind.

Many handsome buildings that were constructed during Colonel Light's time are still standing today, proving that even heavy salt loads are not a death sentence for cities. Adelaide has been dealing with salt for more than 150 years and there is an industry built around urban salinity. The people of Adelaide have learned to live with their salt by preventive action.

PIPING IN A PROBLEM

What people in Adelaide can't control is the salt that enters their house every day of the year through their pipes. In good years about half of the city's water supply comes from the Murray River but in bad years it can be as high as 90 per cent. As the salt load builds in the Murray it comes down the water pipes and into the hot water systems, irons, dishwashers, washing machines, evaporative air conditioners, car radiators and kettles. The salt not only hacks away at these appliances but adds increased costs to households because they have to spend more on soaps, softeners, detergents and conditioners. The CSIRO estimates that the effects of saline water costs Adelaide $55–$65 million per year. At an industrial level some companies have installed desalination or purification plants to remove salt before it adds to manufacturing and maintenance costs.

Don Blackmore from the Murray–Darling Basin Commission says he hates to convert Adelaide into economics but when he does, it makes a powerful statement about priorities in the Murray–Darling Basin: 'The economic performance of Adelaide is about $16.4 billion depending on how you cut that pie. And they use only 1 per cent of the basin's water to support that. They are at least twice as big as any other agricultural industry; therefore they are the economic engine of the basin.'

There is another startling equation about Adelaide's water. Studies by CSIRO suggest the amount of water pulled from the Murray River in an average year to water the city is roughly the equivalent of its annual rainfall, which mostly runs off into Gulf St Vincent. It isn't as if South Australians need reminding of this. Adelaide water jokes are no longer regarded as funny in the city where tens of thousands of homes have tanks at the end of their roof gutters in order to collect the rainwater for drinking. In recent times there have been other innovations including using underground aquifers to store water.

The City of Salisbury, about 25 kilometres north of the Adelaide CBD, has become a water harvester. City engineers have created more than 30 wetlands that serve as storage bays for stormwater that would have otherwise surged straight into the gulf. As the water is slowed down, the heavy metals sink into sediments, reeds filter nutrients and other pollutants are neutralised by sunlight and oxygen. Once the water has been purified it is pumped down into deep aquifers and stored for future use. When summer comes it is pumped back up and used. The free water saves the city millions of dollars in water bills and reduces the demand on the Murray River. Aquifer storage and recovery sites now dot the Fleurieu Peninsula from Willunga in the south to Parafield in the north. Another project has resulted in treated wastewater being piped to the prosperous market gardens of Virginia and Two Wells to the north of the city. The availability of recycled water has allowed the area to double its production, adding as much as $100 million at the farm gate. The uncertain quality of Adelaide's water supply has led to some brilliant thinking and every litre of rain that can be collected leaves more in the Murray to dilute the salt and restore the health of the struggling waterway.

A PUB WITH NO DRAIN

It's not unusual for an Australian problem to be identified in a pub. It just isn't often that it occurs right under the feet of those wise heads propping up the front bar. In Western Australia pub basements became one of the first places indicating that urban salinity was on its way. Not so much a canary in a coalmine as a salt deposit in a cellar. As watertables rose the cellars first became musty and damp and then eventually a publican would open the cellar door and be able to look at his own reflection in a new underground swimming pool.

'Anyone with a cellar knows about it,' says hydrogeologist Louise Hopgood sitting in one of the laboratories in the impressive new

Western Australian Department of Agriculture facility at Katanning. 'The pubs in country towns in Western Australia are pumping out their cellars all the time. Here at Katanning there is a big problem with the swimming pool. They have to keep it full of water because the watertable is so close to the surface that if you empty it in winter it will cause a structural failure. Quite a lot of other towns have that problem with their swimming pools as well.' Originally from Adelaide, Louise is in Western Australia as one of a team of people watching what is going on below the surface of the state's southwest and advising cities and towns on how to cope with the consequences. Katanning is a good example of the scale of the challenge. It is a recurring theme in salinity that the cause and effect are often so far apart geographically as to seem like separate issues. In the same way a rogue irrigator in Queensland causes grief for a grape grower in Mildura, the publican in Western Australia is pumping out his cellar because of forces beyond his direct influence.

Katanning is a four-hour drive from Perth. It is in the heart of the state's wheat and wool country, the area that drives the Western Australian agricultural economy. The town is laid out systematically and well-maintained Federation buildings add to the handsome streetscapes. Like so many other rural centres, it was built in low-lying areas to help bring the railway in. At the historic society building there are piles of photos celebrating the area's farming success. The images show teams of horses dragging carts groaning with monstrous hay bales, long stands of shearers with merino rams wedged between their knees, and a group of men hanging off an early diesel tractor that has cleared a road through the bush. The town still has the largest sheep saleyards outside the metropolitan area.

All that land clearing and ancient salt means Katanning is also fighting off salinity. Louise Hopgood's business is to track the salt and that begins with the soil type. 'We've got about 20 metres of highly weathered granite which is weathered into tight clay down to a very hard granite basement about 20 to 25 metres below the ground surface. At the surface there are more stream-transported type sediments, which are much younger than granite.'

Having established the bad news that the soil is heavy and given to waterlogging and salt retention, Louise then needed to find the location of the watertable. She points out the window to a piezometer next to the main road. 'Through the larger part of Katanning the watertable is actually within 2 metres of the ground surface but as we move higher up to more hilly parts, it can be up to 8 or 9 metres to the watertable. It's pretty much semi-saturated all the way up from the granite basement to within a metre or two of the ground surface.'

With the crucial 2-metre gap breached, the city inevitably displays symptoms like cracking roads and buildings. Near the public swimming pool, flakes of white salt can be flicked off the red-brick changerooms. Across the road from it, a brick fence is reduced almost to a loose collection of worn-out stones. In the pool carpark council workers water a stand of salt-resistant eucalypts that have been planted as a show of resistance—it might as well be a sand wall built to stop an incoming tide.

The watertable rise around Katanning has very little to do with the habits of the good people who live there. They are paying a price for agricultural success. The dryland salinity hurting farms and native vegetation is creating problems for their roads, sheds, houses and public buildings. In 1999 six towns in Western Australia were tested for rising watertables. The next year there were 23 under review and 10 more were added in 2002. The bath is filling up and the towns and cities on top are in trouble. They are caught in a problem they didn't create, can't possibly solve and, which to manage, will bleed them dry financially.

'It's quite a problem in towns because you've got so much money invested in housing and infrastructure,' says Louise Hopgood. 'We really are at the point where there is not much that can be done other than some engineering solutions. At Katanning we are looking at trying to pump out saline water from underneath the town. After that you have to look at the economic side of that system. It may well be that to pump out bores (to lower the watertable) actually costs more than just a little bit of maintenance. It just depends on how much the town is likely to be affected.'

CHAPTER 5 **WHEN SALT COMES TO TOWN**

In 1997 the Western Australian Department of Agriculture initiated the Rural Towns Program to help communities understand their salinity problem, be able to forecast where it will impact and craft solutions. Funding from state and federal governments helps ease some of the bills but towns have limited abilities to raise capital. Katanning has fewer than 5000 people and as salinity costs grow how can they possibly cover the costs of repairs and interceptions? One single bore costs in the order of $10 000 and they need a lot more than one. Then, where should the groundwater that is pumped out be stored? It has to be far enough away to avoid it seeping back into the watertable; otherwise it is like baling out a boat without fixing the hole in the bottom. Even if Katanning grew an urban forest to soak up groundwater it would still be at the mercy of the surrounding catchment that remains nude. For most towns it is a balancing act between finding solutions and repairing the damage. The commitment of locals in the west to deal with urban salinity is another chapter in the folklore of regional Australia—and one that's being carried out to the gentle throb of a pump decanting the pub cellar.

CASE STUDY: CORRIGIN, WESTERN AUSTRALIA

A TOWN OF WINDMILLS LOOKS BACK FOR ANSWERS

Windmills have long provided an image of security in rural Australia and a new side to their value has been reinforced at Corrigin, a town of 1300 residents in Western Australia's wheatbelt region.

Until scheme water from Mundaring Weir was connected for the first time in 1961, Corrigin's essential water supply infrastructure included about 50 windmills and wells that supplied water from deep below ground. However, like in so many towns, the introduction of unlimited water on tap brought rapid changes to residents' habits. Within 10 years only two windmills were still in use, while invisible watertables were

Corrigan Shire president and Rural Towns Program committee member David Abe with one of the production bores reducing watertables in the town.

rising through the combination of extra water for lawns and gardens and less drawdown.

By the mid-1990s alarm bells had begun to ring: waterlogging was showing up in the southwest of the town; the hotel had to pump water out of its cellar; and small areas of salt about 1 or 2 square metres in size were appearing in the business district in late summer. Shire president and local farmer David Abe remembers it well: 'In the early 1990s council organised for a number of piezometers to be put in around the town. By early 1996 they were indicating that the watertables were only 1.5 to 2 metres from the surface in some places. A few of us recognised the problem and we called a public meeting to work out what could be done. A committee was formed that night, but it was hard to know what to do.'

The Western Australian Department of Agriculture in Northam provided some help but a long-term strategy was needed to solve the problem. When the Western Australian government offered to help 13 towns with problems through the Salinity Action Plan from 1996 to 1997, Corrigin made sure it was one of the first to raise its hand.

Understanding how to cope with the town's salinity has grown substantially since 1996, but many agree it has been a steep learning curve. 'Trees, offered as the primary tool, achieved little although some were planted in the first year,' David Abe says. 'Airborne geophysics fly-overs also gave little result.'

But appreciating that excess water has to be pumped out and used on a sufficient scale is now helping Corrigin overcome its problems. 'Through the Rural Towns Program, we have installed more piezometers to monitor the water levels,' David says. 'We are very fortunate that the water is good quality and eight bores are now pumping non-stop into a 182 000 litre tank.' The water cannot be sold by the council but is available for farmers to use for stock and spraying, and is used for irrigating council and school ovals. Exporting water from the town is the ideal option and one which no other town in Western Australia has achieved.

A water truck loads up with free water at the Corrigan stand pipe— exporting the problem out of the town site area and assisting nearby landowners.

'A major cost is roadworks,' David Abe says. 'Where the watertable is high—within 1.5 metres of the surface—we can't compact the ground properly and the foundations collapse. This is one of the council's biggest costs but one we hope to reduce from now on.'

WA Rural Towns Program manager Mark Pridham says Corrigin was one of the lucky towns because of its unique position in being able to take the source of the problem—excess groundwater—and turn it into a resource. This was possible because of the good quality of the water under the town and its location away from flat valley floors.

'When Corrigin was the town of windmills, each windmill pumped about 2500 to 5000 litres per day, removing about 170 000 litres or 170 cubic metres per day for drinking, washing and vegetable and fruit tree irrigation,' Mark says. 'But our analysis shows that once the windmills were removed, the level of the aquifer below Corrigin began rising at about 0.3 metres per annum. It was only a matter of time before problems began in the lower parts of the town.'

Although the total cost of operating and maintaining pumping facilities has been estimated at $22 000 per year, the potential savings in purchased water to the community are about $38 000 per year at current costs. This is making it comparatively easy for Corrigin to make the decision to invest in groundwater pumping as its best salinity management option—and turns a liability into an asset. ■

BETWEEN THE CITY AND THE TOWN

Sitting in his Canberra office with a large map of the Murray–Darling Basin pinned up behind him, Don Blackmore gently waves his hand defensively when asked about the future costs of salinity on the rivers. He has a more immediate invoice. He believes the biggest single cost of salt in the basin in the next 30 years will be on roads.

'Most of our roads have not been designed to cope with high watertables and wet subgrades. So as watertables rise you will see roads breaking up.' Perhaps detecting a raised eyebrow of challenge to his assertion, he smiles and offers an invitation: 'I can take you to hundreds of kilometres of roads now where that's the case—and that is where we are going to pay the price.'

Don Blackmore is far too busy running the Murray–Darling Basin Commission to be taken up on his offer, and in reality it's easy enough to see for yourself where salinity is affecting roads. From Canberra, drive northeast along the Barton Highway, across the Hume Highway and up through the Olympic Way towards Bathurst. During that three-hour run you can't miss the potholes, stressed bitumen and cracks bleeding salt. This is no reflection on the work of the New South Wales Roads and Traffic Authority because when there is a problem the roads are repaired—but a high watertable equals high maintenance.

'With bitumen it doesn't matter if it is salt water or fresh water,' explains Hugh Middlemis from Engineers Australia. 'If you have water ponded close under or touching bitumen it will break it up. That's why you have potholes in roads.'

A road map of the Murray–Darling Basin shows a region covered like a spiderweb with highways and roads. When you multiply those thousands of kilometres of blacktop with Greg Bugden's figure of up to a million dollars per kilometre to repair, then you can see why Don Blackmore confidently makes the prediction he does. In the same way Katanning can't reduce its salinity because of its location, how can engineers design a road to cope with a high watertable? In one section of the Olympic Way outside Cowra, the road slices through green pastures that feed prime beef cattle. As the beasts near the roadside fence they find themselves slopping through ankle-deep water. There is so much recharge from the surrounding land and hills that in the lower parts the watertable is above the surface of the ground. The bitumen sits less than half a metre above it and will quickly fall apart and have to be repaired—until it is inevitably damaged again. Short of building a bridge, what else can be done?

Roads are only part of the infrastructure story. Apply the conditions of dryland salinity to a railway line and see what you end up with. What about a steel shed? Concrete slabs? Fences? Silos?

'It's almost like infrastructure cancer due to too much water and salt,' says Hugh Middlemis. 'The fundamental problem is that when these things were built the watertable was much lower so they were built with materials that were suited to that condition. As watertables have risen, water is coming into contact with those materials and they are starting to break down. For example, if you were going to put a concrete foundation in an area with a high watertable you could use a different kind of cement. The old materials aren't suited to withstand the present conditions. You need a new design and often new materials and it's sometimes a little bit more expensive.'

While 'back of the envelope' calculations can be easily made about the costs of salinity on roads, there is a potential cost that is almost impossible to forecast—flooding.

CASE STUDY: MERREDIN, WESTERN AUSTRALIA

MERREDIN TACKLES THE SALINITY THREAT

Merredin is typical of many towns in Western Australia's broad wheatbelt valleys that were built astride the railway line during early settlement. A century on, the shire retains only 11 per cent of its original vegetation and its groundwater is rising by 10 to 20 centimetres each year. Private and public assets will crumble slowly if groundwater continues to rise without some intervention. Rising groundwater has been a focus of community activity since the mid-1980s and Merredin was one of the first towns to join the Rural Towns Program, initiated under the Western Australian State Salinity Action Plan in 1997. More recently, it was one of six towns subjected to a detailed economic impact study.

Martin Morris, Merredin Shire President and Shahram Sharafi,
hydrologist with the Department of Agriculture Rural Towns Program,
checking progress on evaporation ponds being built as part of the
'Merredin Townsite Groundwater Pumping and Desalination Pilot
Project'.

The good news from consultants URS Australia Pty Ltd is that
Merredin has a breathing space in which to take action. If estimates
are correct, it will be about 2025 before the major impact hits—
time for the town to rally its defences and budget for them.
Consultant Don Burnside says in many cases people see damage
but don't know what to do: 'They carry out some limited action, or
hope it's not happening. Individual residents often live with damage
and patch up cracks just before sale. I know—I've done it myself!
But a year or two later, the problems will be obvious again.'

After extensive consultation, URS was able to apportion costs of
salinity damage to various types of infrastructure, but admits that
they are 'best estimates'. Predictions include $6000 per brick
house three years after groundwater reaches within 0.5 metres of
the surface to pay for building perimeter drains or other means to
discharge groundwater. They estimate a further $2000 to repair
fretting brickwork and crumbling mortar in the first year. Owners
of houses on stumps can expect to pay $1000 every five years for
jacking restumping.

Monitoring groundwater levels in Merredin is an important scientific and educational activity for the local community.

But the big salinity damage costs are repairs and maintenance to roads. Best estimates based on national experience suggest that additional road maintenance due to very shallow saline watertables will be up to $195 000 per kilometre every three years for highways, and $100 000 for local sealed roads within towns.

Applying these principles in Merredin provided a total management bill of about $384 000 in today's dollars using a discount rate of 7 per cent for covering the cost of repairs over the next 60 years. This equals about $130 per resident. Alternatively, early implementation of technologies to control groundwater rise would have a present value of more than $4.5 million. Groundwater modelling undertaken by the Department of Agriculture suggests that about a third of the town may have groundwater within 1.5 metres of the surface in about 20 years, but most damage will occur in two localised areas where water will be within 0.5 metres.

'The investigations concluded that the main reason for rising groundwater under the townsite was due to excessive recharge in the town itself,' Rural Towns Program manager Mark Pridham said. 'Therefore proposed actions will reduce recharge, and should reduce groundwater levels or maintain them at a safe depth.'

A program to reduce imported water, particularly that used to maintain gardens unsuitable for the climate is recommended. Advice to householders about better stormwater management, improved water management on town ovals, and tree planting on vacant areas is also suggested.

Evaluation of pumping to evaporation ponds was another suggestion that has already begun with a grant from the State Salinity Council for a pilot project. Last November [2001] two bores in the central business district began pumping saline groundwater into new evaporation ponds west of the town. A plant has also begun to desalinate part of that recovered water.

Mark Pridham believes that, if proved viable, the technology could be suitable as a combined salinity management and water supply measure for at least 20 other towns in Western Australia alone. While some technical difficulties have been encountered in both pumping and desalination, the good news is that watertables up to 50 metres away from the bores have been lowered from 2.5 metres to 7 metres below ground level. But once pumps were shut down, water returned to previous levels very quickly. Analysis is also showing that the water is very high in some minerals, so may not be as suitable for aquaculture as originally hoped. ■

Martin Morris and Juana Roe, Rural Towns Program and Merredin Water Action Group, inspect damage to historical Cummins Theatre in the main street.

THE BIG WET

On 15 March 1999, as the wet season was nearing an end in northern Australia, a low-pressure system began forming in the Arafura Sea north of Darwin. The monsoon trough drifted west into the Timor Sea where it gathered strength. Three days later the Tropical Cyclone Warning Centre in Darwin named it Cyclone Vance. Soon after, it started swirling west–southwest into the Indian Ocean off the Kimberley coast where it was upgraded to category 3—a severe tropical cyclone. By the time the Tropical Cyclone Warning Centre in Perth took over responsibility for tracking Vance, it was clear this was an ugly piece of weather. By 20 March it was upgraded to category 5. As it intensified it started heading southwest and on the morning of 22 March 1999 it hammered into the continent via Exmouth Gulf. As it arrived it introduced Australia to the most extreme winds ever measured—267 kilometres per hour. The fury of the cyclone was greater than that of Cyclone Tracy, which had flattened Darwin in 1974 and brought with it huge seas and pounding rain. Exmouth was just 25 kilometres from the eye of the cyclone and the town was smashed. Vance then moved inland and started to finally calm down. Before blowing its last gale-force blast over parts of South Australia and Victoria on 24 March, there were record rainfalls near Kalgoorlie and roads and rail to the east were cut off.

Although Vance was the most spectacular cyclone seen in Australia since Tracy, it was the second part in an act that had begun days earlier with a milder, but still nasty, weather system known as Elaine. As Vance was still brewing up north, Elaine was gathering strength in the Indian Ocean almost one thousand kilometres off Port Hedland. Within two days Elaine had been given a category 3 rating. As Vance started to drift southwest, Elaine moved further south and weakened, eventually forming into a rain depression. There was nothing meek and mild about her though, as far as the people of Moora, 150 kilometres north of Perth, were concerned. Elaine dumped more than a year's worth of rain there in only a few hours. The Moore River burst its banks and the town was deluged. Most of the 600 residents were evacuated, having time only to grab some photo albums or as one woman told the ABC, 'my kids and a two-litre milk out of the fridge'. Another couple said that the water came through their house so fast, they woke up to find their bed floating. In the days after the flood as the waters started receding from their destroyed homes and farms, the people of Moora found little comfort in discovering that their storm was not the fiercest ever seen in this country—that was just coming ashore to their north.

In the months that followed the initial flood there were two more floods that surged through Moora as the result of Vance, adding further misery to the small farming community. The flood is commemorated today in a mural painted on a wall in the main street; otherwise the town shows no visible scars of the drama. The mural depicts the community spirit in restoring the town. Outside the town, sheep and cereal crops continue to be farmed, and on Australia Day 2001, Moora's courage was acknowledged when it won the Australian Community of the Year Award. Although the floods were considered a weather event that could probably happen only once every 250 years, options to mitigate the effects of floods have been designed to avoid similar scenes in the future. Part of the planning review involved Mark Pridham from the Western Australian Salinity Action Plan's Rural Towns Program. He believed measurements needed to be made to test whether or not the floods had affected salinity in the area.

'There are complex and numerous interactions between salinity and flooding and each factor has a compound effect on the other,' he explained. 'Groundwater recharge produced by floodwaters increases watertable levels and hence salinity. Higher watertables and increased salinity produces more runoff from waterlogged and salt-affected areas in the catchment.' Piezometers showed that 18 months after the flood, the watertable at most monitoring sites around Moora was between 2 and 4 metres deep. It also suggested that while the levels were not rising, the groundwater pressures were increasing.

There is no doubt Moora suffered a freak rain event but the concern is that Australian regional cities and towns will become more vulnerable to floods for the very same reason they are more vulnerable to salt. Hydrologists point out that the same conditions that cause dryland salinity can also enhance flooding. 'Where the

VIEWPOINT
Hugh Middlemis

HUGH MIDDLEMIS IS A CIVIL ENGINEER FROM ENGINEERS AUSTRALIA WITH 21 YEARS' EXPERIENCE IN HYDROLOGY AND HYDROGEOLOGY. HE HAS WORKED ON WATER PROJECTS ACROSS AUSTRALIA, AFRICA, SOUTH AMERICA, INDONESIA, GREAT BRITAIN AND THE UNITED STATES.

"" *On a broad scale, current estimates are that it will take maybe between 100 and 200 years for Australia's hydrology to balance again.*

We understand the biophysical factors that cause the imbalance. In many cases we have management solutions proposed and in a few cases solutions that have worked. But we have some very good ideas and we are improving all the time. There is more work and more money being spent on applications and approaches to solutions rather than only studying the problem.

In certain situations we need to apply engineering solutions but they can't be applied wholesale. It has to be suited to the site and it has to have the endorsement of the community because change won't occur without the community becoming involved. So while scientists and engineers might come up with some scientific solutions, the shape of the final management solution will probably be slightly different because of the involvement of the community.

Engineering solutions are essentially based on drainage—this might involve passive drainage into a drain or active drainage by pumping water out into a disposal basin. They are expensive to build so they are really only useful in areas of high land-values or high environmental values—where you want to protect a national park for example. Engineers, with the help of scientists, can define the scope of the problem and therefore the size of the drainage scheme to put in. They can implement the scheme and optimise conditions to make it very efficient.

You hear people say they don't think we should be spending money on scientific research to understand the problem. They believe we should be spending it on the ground. Then you hear people say we need to still study the problem because we haven't come up with the universal formula that can be applied everywhere. So although on a broad scale there is a similar underlying problem, every catchment is just that little bit different. So I am not sure if we have demonstrated to the community that we have got that balance right—between spending money on the ground and spending money understanding the problem.

On one hand I am optimistic about our future. I think with further engagement of everybody—community, scientists and engineers—then we can get there in the end. I am also pessimistic because it needs a lot of money. My personal view is that maybe we should revive the debate about the need for some sort of salinity tax. Salinity is caused because of Australia's demand for land clearing and the development of agriculture over the last 100 years, so all of the community is partly to blame and all the community should share some of that burden.

watertable has risen to the surface and where you have essentially saturated ground very close to the surface, it can't absorb any more,' says Hugh Middlemis. 'When it does rain, you have much more runoff so the flooding problem can be significantly enhanced due to dryland salinity.'

THE COST

The size of salinity damage to our continent is overwhelming. Urban salinity's atrophying creep may only be visible as an annoying water pipe rusting in an Adelaide home, a headstone collapsing in Narrogin in Western Australia or a pothole jarring a drive along a road near Goulburn in New South Wales but the sum of its damaging impact is enormous. The Australian Dryland Salinity Assessment in 2000 measured 1600 kilometres of rail, 19 900 kilometres of roads and 68 towns that were at high risk from shallow watertables or had a high salinity hazard. It forecast that in 20 years those numbers would rise to 2060 kilometres of rail, 26 600 kilometres of roads and 125 towns. By 2050, the numbers are estimated to be as high as 5100 kilometres of rail, 67 400 kilometres of road and 219 towns. Toss in Sydney, Adelaide and Perth with significant urban salinity problems and there is a fair part of the national bill estimated at $1 billion per year by the end of the century.

In the history of salinity in Australia there is a small place reserved for the day when a backhoe carved a divot out of the Wagga Wagga showgrounds and found a small inland sea. The best part of the story is that the city responded positively despite a temptation to hide from the problem. 'Very few local government people would come to Wagga because of the stigma of involving their town with the taint of salinity,' says Greg Bugden as he reflects on the journey his city has travelled. 'Today I think local government across New South Wales has its head out of the sand and has looked for direction as to where to go with this problem.' Bob Smith from the New South Wales Land and Water Conservation Department describes it as an 'insidious problem that is actually going through every facet of our life'. And in salinity, as in life, there are as many solutions as there are problems.

CHAPTER 6
Trees and Pipes

*'The desalination of ocean water is even
more important than space exploration.'*

JOHN F KENNEDY, APRIL 1961, IN RESPONSE TO NEWS
THAT SOVIET YURI GAGARIN HAD BECOME THE FIRST MAN
TO ORBIT THE EARTH.

L ike a successful weight-loss client enthusing others to join the
program, Robert Nixon has an impressive set of before-and-after
photographs. The first snap shows him wearing rubber boots and
standing ankle-deep in what appears to be a shallow wetland. The
foreground is covered in halophytes (salt-tolerant plants) and behind
him are dead trees and a greyish landscape. The second image shows
him in roughly the same position but he is standing on firm ground
and the landscape around him shows signs of recovery. Robert Nixon
farms 7000 hectares of sheep and grain at Kalannie northeast of
Perth in the Western Australian wheatbelt. He points across a field of
lush young wheat plants to a stand of gum trees, 'See that?' he asks.
'See what?' 'Exactly,' he grins. 'You can't see the homestead anymore
because of the regrowth. The canopy has even come back so much
you can't see the aerials on the roof and they used to be clear as day.'

The history of Robert's property sounds like it is straight from a
salinity textbook. Land clearing began here in 1925 and the farm was
created out of the prime land that lay at the bottom of the valley.
Robert's father and uncle first started seeing signs of salinity in the

1960s. Now his groundwater is more saline than the Indian Ocean and a third of his property is at risk of being lost to salt. Robert knew he had to reduce his watertable and he began by slowing the recharge with strategic tree and crop planting on the valley slopes. However the groundwater kept rising and the salt was biting into his production. Like a sailor on a sinking ship—it was time to man the pumps.

Robert drilled his first bore in 1995 and started pumping. The site selected was the most active saline discharge and closest to his most valuable assets—the homestead and sheds. However that was just the beginning. Soon the Nixon property had more than 50 bores of which three ran non-stop. The saline water was pumped through more than 10 kilometres of water lines to a series of salt lakes. As part of his long-term plan to harvest salt for commercial use Robert has built an evaporator and crystalliser on site. 'We are pumping 200 tonnes of water every day—that's the equivalent of 43 road trains of water leaving the ground every day.'

Robert is well aware that his process could be seen as transferring a salt problem from one place to another, which is why he talks about salt as if it were nuclear waste. He is at pains to point out that you need to have a safe disposal site for saline water.

It took three-quarters of a century to create dryland salinity at Kalannie and Robert has been pumping for less than a decade. It is costing him money in power and maintenance but his reward is seeing 200 hectares of land coming back to full health. 'We are seeing better germination and better crop production but that's a progressive thing. I think the first thing we always have to bear in mind is we have to stop the process getting worse.'

A RADICAL CHANGE

The question of how to attack salinity is, by nature of the problem, different in every paddock, catchment and city in Australia. What works in one place will not be effective in another. In one area the answer is obvious while in another it requires more science. Sometimes it is cheap, most times it isn't. There is no magic bullet that will kill the salt monster.

Warren Wood from Oxford University is one of the few people who believe it is simple: 'The problem is not intractable from a scientific standpoint. I think it's going to take some bold policy movements on the part of politicians to solve it. It involves land-use practices and that's difficult. I think there are site-specific things where more work is necessary but conceptually it is very straightforward. It is just a question of lowering the watertable. It can be done, there's no question. Make a commitment to do it if you want to do it, then do it.'

So how can a century of absorbed rainfall be reversed? Could Robert Nixon's success be replicated across the board? Tom Hatton from CSIRO in Perth nods his head when asked about the ability of such engineering to reduce the grip of salinity: 'There is no question the problem can be addressed with engineering. The question is how expensive it would be and what are some of the off-site costs that other people would have to bear—those are large uncertainties. You can do "back of the envelope" calculations and say perhaps if we had 2000 or 200 000 or 2 million groundwater pumps across the wheatbelt we would have the problem under control. Those are big numbers and that's the disposal of a lot of salty water.

'If we want to effectively fix the salinity problem we would have to change it [the land] radically and that doesn't mean just a few trees in the back paddock and down by the creek. It means widespread reforestation and that would have profound sociological and economic impacts on the wheatbelt of Western Australia, and not all of it good. If we want to try to adapt to a saline environment, which I think we are going to have to do, that will also involve some profound changes. We will have to learn how to manage saline land and water profitably and we are going to have to learn new ways to value the biodiversity we have left. It's going to take a big change in thinking.'

Tom Hatton is part of a growing network of scientists, bureaucrats, academics and landholders who vigorously exchange information. In the mid-1990s the National Dryland Salinity Program was launched to formalise this network and to spread new knowledge and solutions. 'I think we really are seeing the beginning of a genuine national effort to come up with some solutions,' says Tom reassuringly.

VIEWPOINT
Don Blackmore

DON BLACKMORE HAS BEEN CHIEF EXECUTIVE OF THE
MURRAY–DARLING BASIN COMMISSION SINCE 1990. HE HOLDS A
DOCTOR OF SCIENCE DEGREE FROM LA TROBE UNIVERSITY AND IS
DEPUTY CHAIR OF THE COOPERATIVE RESEARCH CENTRE FOR
PLANT-BASED MANAGEMENT OF DRYLAND SALINITY.

*Salinity is a huge problem but it's not terminal.
We can't have an immediate impact that suddenly
turns it around because you are not going to replace 15 billion
trees and as a community we still want the benefit of our
agricultural wealth and we are entitled to that. So where is
the balance? How are we going to manage it? Well firstly we
have to know where the salt is in the landscape because
nature didn't divvy that up in an even way. Then we have to
cut off the energy that drives that salt out. So we have to
surgically apply trees or engineering and change cropping
systems so it sucks up that water. Now that's the challenge
and the art form for the next 20 years.*

We have no farming system that can be applied broadly that will stop the salinity problem. You can have particular farming systems that applied in pockets can be incredibly effective but you can't roll them out over landscapes. So we need to reinvent our farming systems or at least part of our farming systems so they soak up this water.

We know that we can put trees back into the landscape so how do we convert them into wealth? How do we create an industry out of trees that meets multiple goals of salinity control and the production of wood and fibre products? Now that is not beyond the wit of man to create such a process.

Over the next seven or eight years we have got to be smart enough to get those solutions under way. If we can't do that then we have really got ourselves into a bind because we will have done what we can with big engineering schemes and we will be into much more expensive intervention.

This is the hard-work agenda. This is the generational, almost geologic, change process and quite frankly I wouldn't want our report card to mark that we didn't have the foresight, the guts or the determination to tackle this. It's the migratory birds that will pay the price. It's the wetlands that will pay the price—the river systems, remaining biodiversity and the farming infrastructure. It's too big a price so I don't think that any government in Australia will let it slide.

The art form is to create the knowledge and the processes that will get government and communities to invest their effort and trust that investment in any new start up. That's what the Murray–Darling Basin Commission is trying to do to support governments.

NOTHING LIKE A DRAIN

The Coorong is probably best known as the setting for the classic 1970s film *Storm Boy* based on Colin Thiele's novel. The film reflects the incredible, and at times eerie, wilderness of sandhills and wetlands and the humans and animals that live there. The Coorong is the most significant breeding site in the country for pelicans and an internationally listed area for migratory birds. Every summer flocks of birds, some as small as sparrows, fly down from Siberia, Japan and China to escape the northern winter. They forage during the warm months down under and then make their epic return flight north. Amid the exotic travellers are more than 200 domestic species that call the area home.

The Coorong begins at the Murray Mouth and then stretches out into a peninsula of sand more than 140 kilometres long but never more than 3 kilometres wide. The sandhills separate the roaring Southern Ocean on one side and the peaceful lagoons on the other. The water is a mixture from the river, ocean, freshwater soaks and rain. There are two main lagoons: the northern one begins in the Murray delta and has brackish water while the southern one is hypersaline, meaning up to three times saltier than the ocean (90 000 parts per million). This is because the southern lagoon is mostly void of freshwater influences—only in flood events is it flushed—and so evaporation takes its toll. Despite the often-hostile conditions in the Coorong, the surrounding environment not only copes but thrives. Other than when the northern lagoon fills and spills into the southern (an unlikely event given the Murray usually peters out to sea) the only entry point into the southern lagoon is from a waterway called Salt Creek. While the southeast of South Australia is excellent farming country with a reputation for sheep, cattle, cereals and grapes, the natural fall of the coastal plain is northwest not southeast, and so water accumulates easily in the landscape and is very hard to drain off. European farmers first discovered this when they found great marshes and swamps in the 1860s. When they began working the area they not only cleared the land but cut drains through the landscape to shrink the wetlands. Salt Creek, which traditionally sent only a trickle of water into the Coorong, was widened and deepened to help with the drainage. In recent times the drainage system has been revisited as an aid to reducing the dryland salinity that is affecting 260 000 hectares, threatening another 175 000 and costing up to $15 million a year. Now on a perfect summer day when there hasn't been a sniff of rain for weeks, you can sit by the side of a channel and watch water pouring out of the landscape, down towards Salt Creek and eventually into the Coorong's southern lagoon. The landowners of the southeast have a luxury that farmers in Western Australia can only look upon with envy. They have managed the most straightforward way of engineering a salinity solution—dig a big drain out to sea and put the salt back where it came from a few hundred thousand years before.

PROTECTING THE TALL POPPIES

Driving around the back roads of Tasmania's northeast during summer can earn you a few stern looks. It has nothing to do with the manners of the locals so much as their concern about strangers lurking during the poppy harvest. Tasmania is one of the largest growers of illicit drugs in the world. More than 10 000 hectares of land is under cultivation by 700 growers operating under strict security—especially during the harvest. The opium grown there supplies all of Australia's and New Zealand's medical needs which accounts for about 10 per cent of the harvest. The rest is exported, earning the nation roughly $50 million per year.

Greg Pinkard from the Tasmanian Department of Primary Industry naturally takes pride in the state's farming. When it comes to salinity he uses jargon like 'bottom-up solutions' and 'ownership of the problem' to illustrate how he thinks it is being managed. Standing on a hilltop looking down over a lush valley outside Launceston, he admits that the fairly deep sediment in tertiary basins beneath his feet can lull Tasmanians into a false sense of security: 'There are fairly large areas of salt storage underneath so any change in the landscape that affects water movement is going to cause a problem. We could say that we are reasonably lucky in terms of salinity in this state. We've only got about 54 000 hectares affected at the moment and, even without changing our land practices, by the year 2020 we are talking only 72 000 hectares. So we are in a good position to adopt a "prevention is better than cure" attitude. We are not dealing with big areas or badly affected areas but we are certainly on the lookout because we have got areas of salinity and we don't want these to become major areas, so prevention is the way we are going about it.'

Tom Dowling's family has been farming a section of the island state for generations. Before the poppies came along in the mid-1960s, the area was centred on wool or meat production. Tom estimates up to 5 per cent of his property is affected by dryland salinity. Unfortunately for him it isn't all in one area where it can be more easily managed, so it's a matter of trying to see the saline-affected areas as separate parcels of land.

Tom has used a mix of anti-salinity measures including fencing off salt-tolerant pastures and using trees to soak up discharge. However with poppies earning so much, it's difficult to find the land for further interceptions. 'The economics of agriculture in this area, with land values that now reflect intensive agriculture, clearly dictates that we can't go planting large tracts of our most arable land with trees. So we've endeavoured to use a physical means in terms of surface drainage, or strategic subsurface drainage, to remove this recharge water prior to it entering the soil profile. By keeping the water on the surface and moving it along the drainage lines the water makes it into the river systems and eventually out to sea as it has historically.' With Tom's drainage the salt level in the water is low enough and the flow rate fast enough to mean the saline water makes it out into the Tasman Sea before it can do any damage.

Tom admits that while growing poppies has put a lot of pressure on the landscape, the profits have provided him with the finance to tackle some of the degradation issues that could threaten the future—a future he believes is very bright.

'I am very positive about agriculture in Tasmania. I don't think we have to be clouded into thinking that salinity is going to beat us.

It is just an issue we deal with and manage the best we can to limit its damage to our business.'

The advantage that landowners in both the southeast of South Australia and the northwest of Tasmania have in common is their geographical access to the ocean. There have been wild plans drawn up to construct large-scale drains to pipe saline groundwater from Australia's farming heartlands out to sea. While they sound terrific in theory, the costs are outrageous and drainage is only the simplest of salinity management strategies.

NATURE'S PUMPS

Not far from the poppy fields of northeast Tasmania is another precious crop. On a private farm there is an experimental forest that has been planted by Private Forests Tasmania (a State government agency), in the hope of halting the spread of dryland salinity. The property is lush after winter but there certainly hasn't been enough rain to account for the large pool of water gathering in a prime paddock. The watertable is filling up and discharging into the lower parts of the landscape meaning the salt cannot be far away. David Bower from Private Forests Tasmania pulls a beanie tight down over his head and pulls on a thick jacket and knee-length rubber boots and trudges off towards a grid of young trees planted in a fenced-off hectare midway down a sloping pasture. Inside the wire, he is searching for the salinity win-win: 'What we are trying to accomplish on this property is to manage groundwater more than anything because the movement of water is what is activating the saline discharges. So we are looking at trees to manage the land for water movement and salinity, plus at the same time looking at maximising the productive use of the land. In the recharge areas the level of salinity within the topsoil is fairly low, so we are using conventional plantation species such as *Pinus radiata* in large belts. However with the discharge areas where the water is actually leaking out of the soil profile, we have to find something that can tolerate high levels of salinity and soil moisture.'

David carefully climbs over the electric fence and examines the young natives that are being bent double like old men in the wind. They are mongrel trees—selectively bred for their endurance and toughness. David and his team have looked at 70 different genuses to gauge how effective they are at mopping up water and how useful they are for timber. They have selected hybrids of New South Wales flooded gums, river red gums (*Eucalyptus camaldulensis*) and Tasmanian blue gums (*Eucalyptus globules*). He explains the science involved: 'The breeding has involved selecting a species that will tolerate wet feet and saline conditions and breeding it with two species selected for their timber qualities. So we are trying to produce a hybrid that will have both tolerance for saline water and produce timber down the track.'

The property where David and his team are working runs away from the coast. Applying the drainage solution that works at another farm 15 minutes' drive away is economically out of the question. It would require kilometres of pipes and large electricity bills to run the pumps. If David Bower's experiment succeeds, it will be another important step in salinity management. The thirst of trees can be nature's pumps and in addition to the trees protecting the property from dryland salinisation their timber could also be used as an added cash crop.

Greg Pinkard believes Tasmania is well placed to passively manage salinity and avoid some of the salt disasters it sees across on the mainland. He is right because across Bass Strait the battle to stop salinity destroying river systems of the Murray–Darling Basin has created the most active engineering intervention since Big Lizzie.

BUYING TIME WITH PIPES

Don Blackmore from the Murray–Darling Basin Commission agrees it would be great to reforest the Murray–Darling Basin. He tells anyone who will listen, from parliamentary committees to journalists, that as much as 10 per cent of the basin's upper rainfall areas should be revegetated. He doesn't much care if it's blue gum, river red gum or lucerne, just something to reduce the water energy that drives salt through the landscape. Tom Hatton from CSIRO Land and Water

Division in Perth speaks about the need to grow trees on a 'heroic' level. However everyone knows that trees take time to grow and that is something the Murray River is running out of. The benchmark for salinity levels in the Murray is determined by measurements taken at Morgan, an old river port about 150 kilometres from Adelaide. This is the place where the river is extracted and pumped to Adelaide, Port Augusta, Whyalla and Woomera. In good times it supplies these urban centres with 40 per cent of their water needs while during droughts it can be as high as 90 per cent. The Murray–Darling Basin Commission has set a target for the water taken at Morgan to be below the World Health Organisation's upper limit recommendation for drinking water of 800 EC units 95 per cent of the time. At the moment this is achieved about 92 per cent of the time. That still means the equivalent of 3000 tonnes of salt flows downstream from Morgan every day—twice the amount that comes into the state just over 100 kilometres upstream at the Victorian border. The current salinity levels can barely be restrained, let alone reduced without a dramatic intervention. Land management changes will work but trees are long term—the short term belongs to pipes.

As you drive out of Waikerie in South Australia's Riverland you pass through neat vineyards and orchards of stone fruit and almonds. There are signs posted along the road not for directions, but to explain the need to watch the watertable. Almost as urgent as neighbourhood watch schemes, the message is to find where the watertable is in your area and keep it down. The reason becomes clear a few kilometres down the road at Ramco Lagoon. While it sounds rather exotic, in reality it's a dumping ground for salty water. Even though orange trees blossom just over the hill, the breezes here send a smell normally associated with the beach.

Sometimes the lagoon has mucky water in it, other times it is a saltpan. Fence posts running across part of it suggest the land was once fertile. Now all that grow here are salt-tolerant halophytes. The old river red gums are only useful for birds looking for hollows to nest in. Waikerie is an important part of the South Australian irrigation industry, which is why near to Ramco Lagoon there is a small steel box sitting on top of a couple of pipes that poke out of

CASE STUDY: CAPELLA, QUEENSLAND

REVEGETATION KEEPS SALINITY AT BAY

Capella is 300 kilometres west of Rockhampton and 800 kilometres northwest of Brisbane on the edge of the Central Highlands.

Thanks to strategic revegetation and adoption of new farming technologies local landholders Murray and Trish Brimblecombe have largely beaten a salinity problem, however it didn't seem that way back in 1985.

'At first we were a bit startled when one of the cultivations on our 3100-hectare property developed a bare patch where crops refused to grow. No other landholders around here had heard of salinity problems to this extent. At the start the degraded patch was the size of two four-wheel drive vehicles, however over about three years it grew to about 2 hectares. We contacted the Department of Natural Resources (DNR) and they did some soil tests and identified it as a salinity problem. Previously black tea-tree had been growing in this area. It wasn't long before surface water began seeping out of the ground. It looked brackish and had a salty taste.

'DNR oversaw a tree-planting operation on mounds of soil, because nothing would grow in the salt-laden black soil. In 1991, approximately 1000 trees, mostly salt-tolerant, were planted in 40-centimetre-high mounds of soil, pushed up at 5-metre intervals on the contour. The young plants were mulched and watered regularly. At the time Queensland was entering a severe drought. This allowed rainfall to leach the salts out of the surface soil, permitting native grasses and young trees to re-establish. Slowly but surely the watertable began to recede. Native grasses began to grow back of their own accord and today we have a good ground cover and a healthy plot of trees, many of them about 7 metres in height. My experience sends the message to all Australian landholders, that no matter what your location you have to be awake to the signs of salinity, high watertables and subsequent management that may be required.

'Before the salinity experience we were leaning towards stubble-mulch farming and starting to zero-till with our cultivations. Now all our country is contoured and zero tilled as much as possible. We have been mindful of leaving stands of vegetation for rocky outcrops, to shade our livestock and shelter wildlife.

'A variety of trees were planted to gauge their effectiveness at lowering the watertable. The trees were healthy and had achieved consistently high growth rates. River red gum had an 80 per cent survival rate, compared to 69 per cent coolibah, 58 per cent blackbutt, 60 per cent Queensland white gum, 38 per cent tipuana and 16 per cent yapunya. There was no evidence of salt damage. Initially some people believed the trees would eventually kill themselves by concentrating salts in the root zone but this hasn't happened. Most tree deaths occurred at the seedling stage soon after planting. Chloride levels of the surface soil have dropped by an average of 95 per cent in both the mounded and inter-row areas. The watertable has dropped allowing rainfall to leach salts lower into the soil profile enabling the return of native grasses such as bluegrass, wiregrass and windmill grass.

'The seepage area has stopped spreading but high chloride levels below 50 centimetres suggest that if the trees were removed, the salinity outbreak would re-occur.' ■

With the assistance of Ken Dixon, Soil Conservationist at the Queensland Department of Natural Resources, Emerald (right), Murray Brimblecombe has largely eliminated his salinity problem.

the sandy, salty earth. It is a pump and represents the next stage in fighting salinity here. Ramco Lagoon can only take a small amount of salty water away from the orchards and vineyards. Decades of irrigation have built the saline watertable up to where it threatens the health of the Murray River itself. Retired Riverland fruit grower Jack Seekamp refers to it as the 'dragon lurking below'. To stop the salty water from discharging into the river, engineers have effectively intercepted its flow and reversed it before it causes damage. The Waikerie Salt Interceptor Scheme does artificially what the river red gums used to naturally—suck the water out of the ground. The groundwater is pumped away from the Murray River to an old grazing area now known as Stockyard Plain Disposal Basin. The bores that go as deep as 100 metres pump around the clock and the salty water spreads across this basin and evaporates. By doing this, a total of 350 tonnes of salt per day is prevented from entering the river. It is estimated the salt collected at Stockyard Plain is safe in this natural basin for several hundred years before it will start to creep back into the landscape. This points to the downside of engineering—it is only a temporary fix.

'Engineering solutions really have bought us time for the last decade, otherwise the Murray River would be in a very parlous state,' says Don Blackmore. 'I think to put it in context we need to look at it like this: along the Murray we have 80 kilometres of bore field, that's 80 kilometres of the river with bores pumping 1100 tonnes of salt away from the river every day, 365 days a year. That costs us as a community $2 million a year in power. If we were to switch those pumps off, the Murray River would be unusable for two, three or four months each year. Now that's reality. That intervention has already occurred and without it we wouldn't have the luxury of looking at land-based solutions.'

Stockyard Plain is helping keep the salt level down in the River Murray. Readings at Morgan back up Don Blackmore's claim that time is being bought. In addition to Waikerie there are 15 other interceptor schemes and each records how much salt is being held back. Buronga in New South Wales restricts 36 500 tonnes per year, Barr Creek in Victoria (the first built in 1968) restricts 58 000 tonnes per year, and Noora in South Australia 44 000 tonnes. The three states involved in such schemes use the statistics to prove they are managing salinity and, effectively, buy themselves a credit against further developments that may lead to increases in river salinity.

Governments aren't the only group pumping on a large scale: there are 200 evaporation ponds in the Murray–Darling Basin. Wakool is a tiny service town about 50 kilometres from Deniliquin on the New South Wales side of the river. To reach the town you drive through Australia's rice-growing heartland. Rice farms are in two main blocks: those fed from the Murrumbidgee stretch from Leeton to Balranald in New South Wales, while those watered by the Murray start outside Swan Hill and stretch to Cobram in Victoria. There have been plenty of questions raised about whether or not rice should be grown in Australia at all—especially in an area with an average (and highly variable) annual rainfall of 400 millimetres. What cannot be questioned is the success of the industry. Of the million tonnes grown 85 per cent is exported, putting it in the top 10 agricultural exports in the nation and returning in excess of $500 million annually. Everyone knows that rice is grown in a waterlogged paddy field and to achieve this farms are supplied water by large open earth canals off the Murrumbidgee and Murray rivers. Some farms are so far from the rivers that it takes more than a week, once the irrigation season begins, for the water to flow from the river to their farm gate. Along the way evaporation and seepage take their toll.

The largest supplier of water from the Murray–Darling Basin rivers is a privatised company called Murray Irrigation. Their motto is 'where water flows our food grows' and they live up to their slogan by delivering water to 2400 farms that are using it to earn $300 million from rice, dairy, vegetables, grain and stock. Murray Irrigation uses more water than the entire allocation for South Australia and more than half is used to grow rice. You wouldn't have to be a hydrologist to figure that a crop that requires constant water would soon lead to waterlogging and then dryland salinity. Imagine how quickly rice plants would wilt if the water they were growing in turned saline.

'They go like this overnight,' says grower John Lawrie holding up his rigid index finger and then letting it fall limp. John Lawrie's dramatic demonstration confirms what the piezometers had been forecasting. The groundwater rose during summer when the crops were growing and then fell during winter when irrigation stopped. Some salt readings were as outrageous as 11 000 EC (6600 mg/l), which is almost above stock tolerance. As more and more areas started showing a watertable between 2 and 4 metres from the surface, Murray Irrigation went to the pumps. Getting the water out was never going to be a problem and 24 wells were drilled. Then came the expensive part as 2000 hectares was set aside to create the largest evaporation area in the Murray–Darling Basin.

Hugh Middlemis was a young engineer working on the project and remembers warmly the response of growers: 'Farmers would come up to me in the pub and say, "Isn't that pumping scheme great; that paddock I had before with a salt scald I can now grow a little bit of wheat there". The area benefiting from that pumping is about 50 000 hectares where watertables dropped significantly even in the two or three years I was there.'

Murray Irrigation reported a constant decline in the watertable over 10 years and the average depth is now 3 metres below the surface. It estimates it locks up 240 000 tonnes of salt every year with its interceptor schemes. This amount of salt seems hard to imagine until you drive out of Wakool, past the rice paddies and dairies and arrive at the edge of an inland sea. Pelicans and other seabirds patrol the water. In summer they need to find other accommodation because the sun and wind evaporate the water leaving a blazing white salt crust that would have broken the heart of any early explorer. What has been achieved with pipes could never have been achieved with trees. The equation is 2000 hectares of salt-dumping land plus the power bills to help 60 000 hectares of farms—but just how long this can be sustained is unknown.

PLAYING GOD

Tony Sharley probably has the best office space in Australia. How many other managers can walk out of their workroom into a state-of-the-art cellar-door restaurant overlooking a vineyard sloping gently down to the Murray River? The rammed-earth building is powered and heated by the sun and cooled by breezes that rustle through 250 hectares of grapevines. Standing on the wooden veranda (or sitting with a drink) you easily become absorbed by the abundant birdlife—from pelicans, grebes and swamp-hens to kites, fairy-wrens and galahs. If you can bear to leave the complex there are walking trails, bird-hides and guides that compare with the best national parks. The vineyards are only a small slice of the property that includes 900 hectares of floodplain and wetlands and 600 hectares of mallee woodland. This is Banrock Station Wine and Wetland Centre near Kingston-on-Murray in South Australia, a place with one of the best mission statements ever written: 'good earth, fine wine'. At Banrock Station they are passionate about the environment but this is no green charity. It is owned by one of the largest winemakers in the nation, BRL–Hardy, and the company makes no bones about branding the wine a 'feel good' drop that gives something to the environment as well as to BRL–Hardy. However there is a salinity story to be told here and it begins with the hiring of Tony Sharley. 'I am not a winemaker or vigneron,' he says as he begins his tour of the station. 'I am an environmentalist who had been working for government for 20 years.'

Banrock Station has a history that's common with others along the river. It was mallee country first used for sheep grazing in the 1850s. As the sheep began destroying the understorey, paddle-steamers and infrastructure demanded the timber. In 1925 Lock 3 was built on the property riverfront as part of the taming of the Murray. Like the other 25 weirs, it raised the level of the river behind it by just over 3 metres. It was the third link in a chain of ponds between Blanchetown and Mildura. Parts of Banrock Station were permanently flooded and in doing so the watertable was raised. By 1992 grazing was no longer viable and BRL–Hardy bought the crown lease on a degraded piece of riverfront property—and Tony Sharley got the call.

Driving through the vineyards towards the river Tony tells the story of the salinisation of Banrock Station: 'Water naturally used to flow through this floodplain in a series of channels and effectively many of those channels have been blocked. The moment those channels are no longer able to flow salt accumulates as a powdery dust on the surface of the floodplain. We have seen changes in the vegetation from trees, grasses and sedges to halophytes [salt-tolerant plants] so there's been a whole shift in the type of habitat in this system. Effectively we have created a "saltscape".' To reinforce his point, he stops the four-wheel drive in a moon-like landscape but for the pigface and samphire growing on it; then continues: 'So the obvious solution is to get water flowing back through the landscape and where we have managed to achieve that we have seen quite remarkable results in the shift back to a grassland and woodland type system.'

The next stop on the tour is where you start to understand Tony's belief in playing God. He proudly shows off a sort of Riverland totem pole—a large steel girder stretching several metres high. Recorded on it are tiny signs to represent flood levels over the years. It's part of the culture of the river to celebrate the legends of floods past. Any decent pub on the Murray has a front bar featuring a replica Murray cod the size of a merino ram and black-and-white photographs of 1956. In that year, both the Murray and the Darling were in flood and the system exploded. Some towns were saved with sandbags, others were abandoned by residents who escaped into boats and paddled past the roofs of their homes and businesses. Ask any long-term resident where it got to in 1956 and they will arch their backs and point to a heavenly location often so far from the river as to be unbelievable. Tony Sharley is no exception. He points to the top of his pole that is planted several kilometres from the river: 'The story we are telling here is about the frequency with which floods occur. We now know that salt has been accumulating on this floodplain since we raised the watertable through the development of Lock 3. We want to move that salt through the landscape but unfortunately the frequency of small to medium floods has declined seriously. Instead of seeing floods on this floodplain once every two years we are seeing it once every 10 years. So the salt that is accumulating is having a greater effect on floodplain vegetation, which basically has to stretch much longer before it gets a drink of fresh water. The effect is a decline in habitat, reduced opportunity for fish to breed and vegetation to germinate. So what we need to do is increase the frequency with which these small to medium floods can pass across the Murray floodplain.'

At Banrock Station they have ordained themselves to create floods artificially. Their irrigation water is drawn through wetlands where it is naturally filtered. Each spring they raise the water level in the wetlands and inundate up to 200 hectares of floodplain. The soils are freshened and Tony says there is a recolonisation of the old salty areas with increased growth in red gums, grasses, sedges and lignin. One of the prominent trees in the original woodland was blackbox and through the work of local scientist and former fruit grower Jack Seekamp a salt-tolerant hybrid has been developed and is planted at Banrock Station. When asked if there is a danger that the salt washed out of Banrock Station floodplains will end up downstream as someone else's problem, Tony responds that salt is not used biologically so if it is kept moving it will eventually flow out to sea. Keeping salt moving seems to be a common thread among those trying to manage salinity. With drains, pipes or artificial floods, the idea is to never let the salt build up and destroy the land and what sits on it.

But how do you manage salinity if you are forced to bring it onto your property?

PROTECTING AGAINST IT

An irony that would bring a bitter grin to the face of any Australian farmer is that irrigators often have to overwater their crops to avoid salinity, and in doing so, create more salinity. Irrigation has been a driving force in increasing dryland salinity. The waterways that drive irrigation are therefore becoming saltier. So every drop used to grow grapes, rice or oranges brings with it a tiny hit of extra salt for the system. The salt then builds up around the root zone of the plants threatening their survival. What is the simple answer to this? Add more water and flush the salt further down below the root zone? Trouble is that the extra salt and water leach down until they join the watertable—and we know the rest of the story from here.

Almost since the first irrigation systems were installed by the Chaffey brothers, subterranean drains have been used to help avoid the problem of excess water. These tile-drains were effectively ceramic half-pipes that caught the water and ran it off either back towards the river or to lower parts of the landscape. They are still the first line of defence on many irrigation properties and terracotta pipes can be seen in piles around new plantings. Again the idea of the drains is to keep the water flowing and the salt moving. Along the Murray and Darling there are dumping grounds that are filled from tile-drain outlets. Those arguing for sustainability see a vicious cycle of water use leading to drainage overuse and mini Lake Eyres dotting the country.

David Dawes is a modest man who doesn't waste words—he doesn't waste water either because he knows if he does it could kill his family property. David's grandfather took up the property, which lies just outside Wentworth in New South Wales at the confluence of Australia's two greatest waterways, the Murray and Darling rivers, at the end of World War II and ran it purely for wheat until the early 1970s. David and his father added cattle to the wheat but by the early 1990s they saw the writing on the wall: 'We could see what we were doing out there wasn't sustainable and long term the chance to better our lifestyle and consolidate the property was limited.'

So David made some bold moves. He and a business partner divided the property in two, selling half to fund their expansion plans. They then planted 320 hectares of chardonnay, shiraz, cabernet and merlot grapevines on the gentle rolling country. They signed a contract with a large winery in Mildura and started growing grapes. Their lifeblood is a steel pipe that runs from their farm gate, under the highway and down to a galvanised-iron pumping shed nestled between two enormous river red gums on the banks of the Darling. This is one of the last licenced irrigators on the river and the water barely flowing past looks like milk. David uses on average 800 megalitres (400 Olympic swimming pools) of water. It costs him about $1000 per megalitre to buy his high security (guaranteed) water which he roughly doubles in value when he sells the grapes. But David gets not only water coming up the pipe—the river in her last gasp before joining the Murray gives quite a salt load along with it.

Environmental regulations mean David can't return any water to the Darling, so what goes to the farm, stays on the farm. This is why David is considered a model for irrigators. 'The salinity of the Darling River that we pump out is around 200 to 300 EC units. That equates to a net import to the property of 200 to 250 tonnes of salt per year with the irrigation water,' he says. 'We knew from soil monitoring there were reasonable salt levels in the soil so a combination of regular soil tests and sap testing [sap juice from the vines] during the growing season, gives us a very close indication of these levels and the build-up in the vines.'

In the style of the old tile-drains, David has laid about 35 kilometres of 4-inch (10-centimetre) drainage pipes beneath the root zones of his vines. When the salt levels in the vines indicate there is a problem they follow the old formula of overirrigating and leach the salt into the drainage system. From there it flows to a 10-hectare site planted with salt-tolerant eucalypts and other natives to mop up the water.

Monitoring almost every drop used means David spends part of each day on the computer in a corner of his machinery shed. His hi-tech approach includes watering his vines with thousands of tiny drippers developed in Israel. These drippers have an elaborate mechanism that keeps the water turbulent so it doesn't get clogged up with the muddy water. It also allows him to precisely deliver only the exact amount of water needed. 'It hasn't been too hard,' says David in reflection. 'The information is out there and it's happening in many areas around Australia. It's just a matter of which technology best suits us here and we seem to have it going quite well now.'

After a lot of science and planning David believes he has protected himself from the annual salt import. Not everyone is as fastidious. It is still common in wine-growing areas to see pivot sprinklers sitting on pipes above the vine canopy shooting jets of water that use more than is needed and load the watertable. It's a shotgun approach compared to David's surgical strikes. A study by the South Australian Department of Agriculture at Loxton showed as little as 20 per cent of water is all that reaches the plants under such irrigation methods. The Australian wine industry boomed in the last part of the twentieth century. In 1991 Australian exports were $174 million and by 1999 that had grown to $1 billion and was still growing. It uses just 3 per cent of all irrigation water in the nation but saline water threatens its future.

Scientists, farmers and environmentalists around the country all agree that because of salinity the quality of our land and water in Australia is deteriorating. Dryland salinity creeps across landscapes slowly degrading them while water carries salt to new places. Given that the drain hole of the Murray–Darling Basin—the Murray Mouth—is almost constantly in danger of closing over, then most of the salt is being kept in the system. Irrigation water will continue to deteriorate in the basin as the salt load builds. How will the cost to irrigation be measured? At what point is it unsustainable to use saline water and where will the fresh water come from?

JFK'S DREAM IS KANGAROO ISLAND'S REALITY

Kangaroo Island lies an hour by ferry from the South Australian mainland. Its isolation makes it a wilderness of significance because few of the many native inhabitants can swim. Tourists from around the world are thrilled at seeing echidnas, wombats, kangaroos, koalas, birds and wildflowers in vast numbers. The island's Flinders Chase National Park alone covers almost a quarter of the island, and the beaches are so clean and the water so clear they once topped a national field to be judged Australia's best. The main ferry service berths at Penneshaw—a tiny town of 200 people set around a bay of perfect white sand and sparkling, clear blue water. About the only thing Penneshaw has ever had a problem with is its water. Although Kangaroo Island has had enough land cleared for farming to have created a dryland salinity problem, for once it wasn't salt that was the villain. The town's original water supply was gravity-fed from two dams on nearby farms, a system that worked well until too much strain was put on it. 'The water was awful,' admits Don Bursill, Director of the Cooperative Research Centre for Water Quality and Treatment in Adelaide. The only solution seemed to be to build a long and expensive pipeline from the reservoir that supplies the island's main town, Kingscote, with water. Instead SA Water Corporation engineers looked out to sea for the answer.

In 1999 a desalination plant was built at Penneshaw inside a shed the size of a tennis court. It sits a stone's throw from the ocean and converts seawater into drinking water. The water is drawn into the plant and forced at high pressure through a membrane that breaks it down to a molecular level, sending the fresh water one way and the salty water back out to sea. For every 100 litres of seawater that runs through the plant, about 45 litres of fresh water is produced. While, initially, there was some concern about the effects on the marine environment of pouring hypersaline water back into the sea, the amounts are so small as to be almost, literally, a drop in the ocean. The drinking water has a salt level of about 200 milligrams per litre (less than the Murray River) and is pumped into large storage tanks and gravity-fed back into the town. The $3.5 million system requires no chemicals and only routine maintenance.

In 2001 the cost of Penneshaw's water was twice as much ($2 for 1000 litres) as elsewhere in South Australia but some believe the Penneshaw solution could become more and more viable in Australia. Hugh Middlemis, from Engineers Australia, thinks in the next 20 years the price of desalination technology will come down to the extent that it will be a very cost-effective means for supplying water to capital cities on the coast. Don Bursill is a fan of the desalinated water and thinks it offers a chance to improve the quality of water inland: 'A lot of rural inland towns have brackish water supplies that are really unsatisfactory. In South Australia we pipe water from the Murray River, or other surface water sources, hundreds of kilometres sometimes at great cost and usually with a

HOW DOES DESALINATION WORK?

Desalination is simply the desalting of water whether it is brackish groundwater or seawater. There are several different ways of achieving this. The most basic system that has been used for hundreds of years is to simply boil the water and collect the steam. The steam contains no salt so when it is condensed it can be used for drinking. Other methods of desalination are:

Flash evaporation is a more sophisticated way of boiling salty water. In this case the water is heated and then pumped into a low-pressure tank. As the water hits the subsequent drop in air pressure there is a flash effect as the water instantly vaporises into steam. It is then collected and condensed. This system is used in some desalination plants in the Middle East to convert seawater into drinking water.

Freezing, the reverse of flash evaporation, can also desalinate water. Fresh water has a higher freezing point than salt water. Using that principle the temperature is lowered to the point where the fresh water freezes but not the salt water. The fresh water forms ice crystals in the brine that can be removed and melted into drinking water.

Electrodialysis is a more complicated filter system that uses an electric current to separate the salt from the water. The dissolved salt splits into positively and negatively charged ions, which are attracted to electrodes with the opposite electric charge. By creating a maze of membranes or cells that allow only one lot of ions to pass through, eventually a stream of salty water is ejected leaving fresh water behind.

Reverse osmosis is becoming the most popular form of desalination. This is basically a filter that is able to separate pure water molecules from all contaminants, including salt. The seawater is forced through at great pressure and then the two water streams are pumped off in different directions.

lot of difficulty to small communities where they have brackish water underground. I really think that these developments on desalination will mean we are going to be able to use these brackish groundwater resources for better quality water for rural communities and at a reasonable cost.'

Don Bursill's idea for inland Australia is already happening at Coober Pedy and Roxby Downs. These two communities in the remote far north of South Australia exist because of lucrative opal, copper and uranium mining returns—so money isn't such an issue.

While the economics have never been right in Australia for desalination on a large scale, it has proven successful in many other countries. The Israelis built the first major coastal system in 1965 when they started distilling seawater to use in Eilat. Arab countries in the region followed suit, as did areas in South Africa. In the United States, Key West in Florida started desalinating seawater in 1967 and has recently built a new reverse-osmosis plant (*see box, How does desalination work? p140*). Land-locked Buckeye, Arizona (population 4000) is a town in the heart of cotton-farming country that uses desalination to convert salty groundwater into a potable product.

FARM WITH THE LOT

When you discuss salinity with farmer Kim Diamond you should expect direct answers and little sentimentality. There is a certain restless energy in him as he gets behind the wheel of his ute and starts driving towards one of his properties north of Perth. Outside, the country is laid out with broad-acre crops broken only by small stands of mallee. 'People come out here looking for salinity and they take a picture of a tree that's been dead for 50 years,' he barks with a tinge of bitterness. 'Don't focus on that stuff, look at what is going on to manage it.'

Kim Diamond has been focusing on ways to overcome salinity ever since he was old enough to start farming full-time. He says he was lucky his father had the patience to allow him to experiment and go away and learn things. He learned enough to expand the one family property into several and along the way he collected a shelf full of awards. He has started Landcare groups, set up seminars, lunched with federal and state politicians, lost count of how many agriculture ministers he has briefed and travelled to just about every corner of the country. 'I am a bit of a sucker for that,' he says. 'If I am on holiday and I see a cocky out in a paddock I have got to stop and talk to him and see how he is going.'

Kim shares the belief of many that the wheatbelt of Western Australia is driving much of the thinking on salinity management. He believes farmers in the northern agricultural areas of the state in particular are totally aware of the problem and what is creating it. For Kim Diamond the search for ways of using the rainfall efficiently never ends. He loves talking about sustainability almost as much as he hates talking about land that salinity has killed. He may refuse to acknowledge it but he knows the danger of salt. In the early 1960s there was no salt-affected land on the property. The family had 800 hectares of crops and up to 4000 sheep. Salt started taking away the best ground and soon they were running less than a quarter of the number of sheep and losing cropping areas every year. Through trial and error, stock numbers are back up and cropping continues. As he brings the ute to a stop near a field of tiny green barley plants he runs a hand through his thick crop of white hair and says slowly and deliberately: 'I don't see any one thing as a stand-alone. I think we have to stop looking at salinity management as stand-alone issues and instead integrate them into whole farm-systems. One without the other is less successful and less economical and probably less sustainable. That's what I am showing you here.' With that he steps out of the vehicle and gives a lesson in total farm salinity management. 'We realise we've got the problem and what we are endeavouring to do is use the problem as a resource rather than something we can overcome entirely.' Kim has created a patchwork system that uses all the salinity management tools to reduce water draining into the watertable. On top of the hills are trees to soak up recharge. In the higher areas the crops are planted on contours, which gives better water surface control and reduces runoff and erosion. Halfway down the paddock is a fenced-off plot of lucerne that absorbs more of the water that is flowing underground. The groundwater at this point isn't very salty so the deep-rooted perennial crops can use it. By the time the water reaches the bottom of the hill it is too salty for most plants so stands of salt-tolerant trees are growing alongside salt-tolerant shrubs such as saltbush and bluebush. Kim points out the machine that he developed and patented that builds raised beds for direct sowing of saltbush into saline ground. Land that was heavily salted and useless for cropping is now covered in saltbush for sheep grazing. Finally at the bottom of the valley, drains are dug 2 to 3 metres deep to carry the excess water away into a natural salt-lake system.

'Our aim is to develop a system that is both economically and environmentally sustainable. We believe with the introduction of the perennials of the salt-tolerant type and the better quality water users up slope we can achieve that,' says Kim, allowing himself a small smile. 'I think the science is in place to assist in the management of these areas. I don't have any doubt about that but I think there is alienation between landholders or end-users of the technology and the people who are developing this stuff. I think scientists, extension people and researchers need to be seen to be working or talking to farmers, not at them.'

Kim Diamond is the sort of landowner others talk about needing more of in Australia. Syd Shea from the University of Notre Dame Australia respects the tradition of farming in his state because both his grandfathers cleared land and farmed in the state's wheatbelt. He believes the ingenuity and determination shown by previous generations will be repeated in the current generation's battle against salinity: 'I look at this land and see what they [farmers] have achieved and the tough times they went through and the successes of the scientists and then I have great confidence. The genes haven't changed. I think that in this new re-engineered agriculture we will capitalise on that same innovation that created this fantastic industry—so I'm optimistic.'

From Tasmania's poppy fields through to the Murray–Darling Basin and out to Western Australia's wheatbelt, variations of pipes and

WHAT IS SO SPECIAL ABOUT LUCERNE?

Whenever the salinity big picture is discussed there is talk of changing agricultural systems and practices. Scientists and farmers are always looking for these new systems and the Holy Grail is to find a crop that can be grown within current farming patterns and that produces a return while reducing salinity—lucerne is one such crop. The legume has been part of cropping almost since humans first began planting seeds. It was first grown in the Middle East and around the Mediterranean before being introduced to North America where it is known as alfalfa.

Lucerne is grown as a fodder plant, so it is harvested like hay and fed to livestock. It is full of proteins, vitamins and minerals but in salinity terms its best feature lies beneath the ground. Lucerne is a perennial so once it is harvested the stubble will reshoot the next season. The plant puts down a root system that can go as deep as 9 metres, which enables it to draw nutrients and water often not available to shallow-rooted crops. This means that not only can it survive in harsher conditions—but it can also drain the watertable and reduce the dangers of capillary rise.

In 1995 CSIRO began a five-year experiment at a property near Katanning in the Western Australian wheatbelt. The idea was to compare the water use of lucerne with that of annual pastures like sub-clover. The study showed lucerne used a greater amount of water and from a greater depth in the soil—the sort of moisture that would have trickled down into the watertable. In addition the study showed that lucerne grew for longer. When the clover came to an end at the beginning of autumn, lucerne kept growing and using more subsoil water. CSIRO scientists measured the clover roots at 50 centimetres and the lucerne at 1.5 metres and estimated that the lucerne sucked an extra 40 millimetres of water from the soil. Another CSIRO study showed that in areas with less than 600 millimetres of annual rainfall lucerne is the equal of large trees as a means of preventing leakage into the watertable.

Other studies have shown lucerne has a slightly higher salt tolerance that other crops and that there are genuine benefits from including lucerne in a crop rotation. The economic return of the reduced salinity risk is implicit but difficult to calculate.

Lucerne has been a more traditional crop in the eastern states but is now being taken up more extensively in Western Australia. Typically it is sold as dry fodder for stock but a future economic driver could come from domestic and market gardeners who are finding it to be a terrific mulch for vegetables and flowerbeds. In Europe there have also been moves to use the protein known as rubisco, which is found in lucerne, as a replacement for soy protein in food for humans.

trees are being used to manage salt. However salt is not something that goes away. An old-man saltbush plant absorbs saline water through its roots and eventually the salt finds its way out onto the leaves of the bush that are nibbled off by sheep and digested. The salt goes through the sheep's system, exits via droppings and ends up back onto the ground where it had been first absorbed. Salt washing down river systems in the Murray–Darling Basin will eventually drain into the Murray River. The Murray Mouth almost closes over every year because so much water is extracted for irrigation there isn't enough water to push the salt out to sea. The water that does flow through the barrages at Goolwa and into the Southern Ocean certainly takes a lot of salt with it but a vast majority stays in the system. Australia has always had poor drainage and things haven't got any better.

In dealing with the silent flood many people talk about salinity management, few talk about solutions. What then can be done with all this salt?

VIEWPOINT
Alex Campbell

ALEX CAMPBELL IS A SHEEP, CATTLE AND AGROFORESTRY FARMER FROM NARRIKUP IN WESTERN AUSTRALIA. HE HAS BEEN CHAIRMAN OF LAND & WATER AUSTRALIA, THE NATIONAL DRYLAND SALINITY PROGRAM AND THE WESTERN AUSTRALIAN STATE SALINITY COUNCIL.

I believe the national mood pertaining to salinity has really changed in recent years. We have stopped thinking of it as a farmer's problem. We have stopped apportioning blame as to who caused it, whether it is governments or farmers and we are now being proactive about it. We are asking what can we do to manage salinity, our watertables and the wider environment.

We live in the driest continent on earth yet we are failing to use our annual rainfall to the best advantage. What we are saying to farmers is if you can find a way of using that surplus rainfall to better effect then you will solve your salinity problem plus your economic and social problems relating to agriculture.

I think we are facing something almost as big as the change from horses to tractors. When you think about that, it wasn't just the farmer who had to change a lifestyle and a way of doing things, it was the blacksmith's shop that had to change and become a garage, it was whole infrastructure that affected the whole community. Now we are almost at that stage again. We need to get from an annual-based agriculture to a perennial-based agriculture which is a huge change and that's the only way we are going to solve this.

Look at the natural bush. It has evolved over millions of years to be in sympathy with Australia's droughts, floods and soil types. So we really need to look at that natural bush and say well how is that coping? Then we need to copy or mimic that in our farming systems.

If I could paint a picture of the future Australian landscape, we of course will have our wheat, barley and sheep but it will be in a mosaic with perennial farming. Those perennials will have two basic forms: there will be the permanent woody perennial like oil mallee fodder shrubs and agroforestry and as well there will be a range of phased systems. These might involve lucerne or some perennial cropping system for a number of years and then change back to an annual system. It might be an integrated and dispersed annual and perennial system that some people call opportunity cropping where you have a summer crop and a winter crop. It will be a real mosaic of annuals and perennials, shrubs, trees, grasses and crops and by Jove it needs to be a better Australia than we have at the moment.

CASE STUDY: EAST CARNAMAH, WESTERN AUSTRALIA

KEEPING FRIENDLY WITH THE LOCALS BRINGS RESULTS

Keith Camac crops wheat, lupins and canola, and grazes sheep for wool on his 6000 hectare property, Daldee. He has found native, rather than exotic, species the key to salt management.

'Salt first appeared along a creek line on this property in the mid-1950s and my father used contouring and many new plants to try and reduce it. But after little success with paspalums, creeping saltbush, clovers and others, we found the local bluebush (*Miireana brevifolia*) to be the best of all, both as a salt-buster and a drought-buster.

'We now have 200 hectares of bluebush over five paddocks and have been using some areas for ewes at lambing for nearly 40 years. Not only is it controlling salinity, but the sheep like it and do well on it.

'In the 1950s and 1960s we tried, unsuccessfully, to revegetate with recommended clovers and veldt grass. In 1972 we scarified and re-fenced the area and seeded with samphire, which wasn't too successful, and bluebush.

'The local bluebush seed was harvested by hand from a small, established patch, scarified and then spread by hand over the rest of the area. It did very well and continued to spread by itself. We have found that bluebush will grow just about anywhere if the ground is ripped to establish it and it is fenced off.

'Another species present in the area was old-man saltbush but it didn't spread as well or as quickly, although now it is slowly spreading through the bluebush.

'Once land becomes bare I defy anyone except the Dutch to get it back to grain production, but reducing the watertables by growing perennials such as bluebush can prevent it getting to that stage.

'Every year we top-dress the bluebush with 50 kilograms of double phosphate. It is normally grazed from March to the break of season. Occasionally there has been a disease that kills it off for a season, but doesn't appear to affect it long-term.

'Because wool prices are still low, we are about two-thirds cropping and one-third grazing. About 4500 hectares is available for cropping wheat, lupins and canola. Normally we run about 5000 sheep through the year including 2300 to 2500 ewes. Without the bluebush, we wouldn't be able to run these numbers.

'Each year we bring the ewes into the bluebush from crop stubble in March or April—first leaving the gates open so they can move between the two areas. They normally have about four months on the bluebush over autumn when there is little else available, then when the season breaks and the pasture has become established, we move sheep out to clover. 'We have also planted between 3000 and 6000 trees around the edges of the bluebush over the last five or six years, mainly salt-tolerant river gums, york gums, *Casuarina obesa* and *Acacia saligna*.

'We always fence the plantings off and use tree guards to protect the young plants from rabbits and birds. This raises the cost to about $2 per tree but survival rates are excellent. According to Dad's records the area started showing signs of salt in the mid-1950s, possibly following the big flood of 1955. Using as much groundwater as possible seems to be the way to control it.' ■

East Carnamah farmer, Keith Camac, in bluebush pasture ready to stock lambing ewes.

CHAPTER 7
MAKING THE
BEST OF IT

Always back the horse of self-interest

JACK LANG

It is said that when Michelangelo was asked how he sculpted such masterpieces, he responded that all he did was wander around quarries looking for statues stuck inside lumps of marble. Once he had found Zeus or David he would then free them by chipping away the bits that were holding them back.

Rodney Holland is no Michelangelo but he is a hero to anyone who has ever looked at a pile of junk and thought, 'I reckon I could do something with that'. A few years ago he bought the waste dump outside the town of Cooltong, near Renmark in South Australia's Riverland. It didn't cost him much and there's a very good reason for that. It was a mess of mallee bush filthy with car bodies, fuel cans, tyres, plastic and rusted farm machinery—and those were the nice bits. After carving out a small patch to live on, Rodney began cleaning the place up. He took so much scrap metal to the recyclers that the returns paid his mortgage. There are still some bits and pieces lying around if you look for them but the biggest trouble he has these days is keeping out the local hoons who still consider the old dump a proxy Mount Panorama when they are full of Bundy and bad manners.

After several years of housekeeping, Rodney is moving forward. He has subdivided a section of the now pristine bush and sold blocks to locals who want to live in a twenty-first century home in a landscape that still looks like Australia in the eighteenth century. The trouble is Rodney can't stop thinking about how to turn waste into want. Part of his property contains the Cooltong evaporation basin. It is a dumping ground for waste water and has been since farmers began putting in fruit crops after the war. To avoid rising watertables and salinity, the farmers buried tile-drains underground, so when they irrigate, the excess water and salt drains away from the roots and down to a central repository. That repository is the Cooltong basin. A groaning sound, which in winter can be heard almost daily and in summer seemingly ever hour, indicates another surge of water is coming through. It pools in a concrete tank and then is pumped through Rodney Holland's backyard and into the basin.

Reportedly some fruit growers have wept when taken to see the basin. Certainly most people are taken aback when they first visit. The dead trees protruding from the water are a deathly grey colour. There is a curious feel to the lake surround. It is soft and spongy underfoot with a texture not unlike insulation batts. The water itself tastes like the sea. Waterbirds give an impression of a productive ecosystem but there is no doubt this is an environmental dead-end. It is, like any other dump in Australian society, a price we are prepared to pay for our lifestyle.

Urban legends are built on stories of weekend renovators who go to a council dump with a trailer full of refuse and return with a treasure someone else was about to offload. Rodney Holland believes the water surging through his property is a treasure: 'I am a bit into recycling anyway and one of the things I thought about was aquaculture because the water is still very usable—it's not contaminated—so one of the things I thought about was yabbies.'

The water that flows from the central irrigation point is mildly salty, about 2000 parts per million, but still usable. It is only when the water arrives in the basin and sits there evaporating, that it becomes too saline to be recycled. So rather than letting the water flow straight through his property to the basin, Rodney experimented by feeding the water through a series of canals he had dug. The water winding through the channels doesn't pond and evaporate and is constantly refreshed with new flows, so the salinity level stays constant. 'It comes in at 2000 [parts per million] but because I am not holding it, the level won't go above 2000 parts, and I'm trying to combat that with saltbush and eventually I'll have trees planted.' Rodney proudly drops to his knees beside a trench and pulls a rope up through the algal skin that grows on top of the water. Eventually a large tunnel-shaped net surfaces and when the lid is sprung open, it reveals bluish-black yabbies crawling over each other. 'If you think about it, look at what grows in seawater. As long as the water isn't contaminated and at this stage it isn't, I think we can reuse this water. I hope it turns out to be a fairly good commercial venture using wastewater. I'd like to see my grandkids benefit from it because it has the potential.'

There is a school of thought in Australia that says why fight salinity? It is too costly, too hard and too widespread. Why not just live with it? Instead of crying about what we can't do, try finding out what we can. The economics alone are a strong argument. The National Farmers Federation and the Australian Conservation Foundation estimated $65 billion needs to be spent over 10 years just to get a handle on things. David Chittleborough from the University of Adelaide thinks that is an underestimation. People like Rodney Holland don't think on a scale like that but they still see a bigger picture. He cannot fix the cause of salt running out of irrigation farms, but he can try to use the saline groundwater for something.

At Wakool in the New South Wales Riverina district, where engineering works pump salty water to the surface to lower the region's watertable, an aquaculture experiment on breeding baby snapper has begun. It is the latest attempt to use saline groundwater as a resource. The first successful effort was a barramundi farm at Grong Grong in New South Wales. Others have also worked: the Northern Territory University has successfully run a hatchery for trochus (a marine snail whose shell is valuable for jewellery and button making), and in Victoria, South Australia and Western Australia scientists and farmers have been able to breed tiger prawns, tuna, black bream, shrimp, salmon, marron and redclaw crayfish. Frustrated scientists who battle to get the right levels of pH, oxygen, nutrients and salt in the water have sneeringly referred to inland aquaculture as a hundred ways to kill fin fish. The breeding of larger fish, such as snapper, seems to be more successful but getting all the levels of the ingredients right and then keeping them consistent is difficult. A successful formula would potentially create a large industry given the amount of saline water being diverted away from rivers not only in the Murray–Darling Basin, but also in parts of South Australia and Western Australia. According to Stuart Fielder from New South Wales Fisheries, if the project is successful and uses just 10 per cent of the Wakool evaporation basins (200 hectares), it would produce the equivalent amount of snapper that is imported each year into Australia.

Aquaculture is Australia's fastest growing primary industry. It has increased by an average 15 per cent every year since 1990 and in 2002 was close to being worth $1 billion. With improvements in technology there seems to be no reason why saline groundwater can't be used instead of seawater. But when it comes to saline water, there is an even more basic harvest—the salt itself.

THE SALT MAKERS

John Ross may have the most unusual pair of occupations in the country. Half his time is spent selling real estate in Melbourne; the other half is spent trying to find an economic way to sell the gourmet salt he makes at his property in an area known as Pyramid Hill in the Tragowell Plains of northwest Victoria. The area was given its name because of the resemblance the local high ground has to the Egyptian landmark. That such a hill would stand out is also indicative of just how flat the country is that surrounds it. Pastures spread as flat and green as a billiard table, but lurking just half a metre underground is some of Australia's most saline groundwater. Property in Pyramid Hill started to go out of production as early as the 1930s and in some areas the groundwater is as salty as the sea. John Ross and his business partner Gavin Privett don't want to keep it underground. They bring the water to the surface into a series of plastic-lined dams that act as evaporation ponds. Exposing the brine to the sun and the wind eventually reduces the contents to a thick crust of salt that is harvested and bagged. The grade of the salt can vary and is used for industry or agriculture. Several years ago a chef asked if they had any flake salt because he preferred to use it in his kitchen. Although at the time they didn't know what it was, John and Gavin did their homework and developed a special stainless-steel oven to make it in. Later they experimented by adding native bush herbs and spices for a more gourmet appeal. Each year roughly 3000 tonnes of salt are harvested with plans in place that are aimed at pushing this figure to 15 000 tonnes. A happy by-product is that the watertable underneath the property has been lowered by 5.5 metres and surrounding farms are finding their watertable is down as well.

John Ross is happy with the salt production but one day hopes the salt water will also indirectly power his plant. The business is experimenting with a solar gradient pond designed to use saline water to harness the sun's power. The 3000-square-metre pond looks like a dam, which has large plastic circles floating on top to keep the surface calm. John explains it works on the principle that different densities of salt water do not mix and will form into layers: 'We have different levels of salt water with different densities. The densest is on the bottom of the pond and when the sun shines the heat goes down through the layers in the top and lodges on the bottom. Each of the layers above it creates what you might term an insulation blanket, with different densities of salt water. We found a biological way to keep the top of our ponds clear from algal growth. We use brine shrimp, which we use extensively in the salt fields to clean up the brine, and this is the first time this has been done in the world that we are aware of. To harness the heat that is in the bottom of the pond, we run pipes containing fresh water through it. The pipes pick up the heat of course and we transpose the heat to our plant.'

The heat stored in the solar pond can rise to 80 degrees Celsius and John believes this type of solar power source will be attractive to remote locations and will become increasingly viable as Australia tries to reach a target of 2 per cent green power by 2010.

Even though evaporation basins across Australia are estimated to create a million tonnes of sodium chloride each year, having a mountain of salt is one thing—moving it to market is quite another. It has to be collected, packaged and marketed. The coastal salt operations are so well organised and on such a scale that breaking their grip on the market is a David versus Goliath proposition.

A BITTERN HARVEST

Salt isn't the only product that can be extracted from saline water. In the final stages of production at Pyramid Salt, a bitter oily by-product is created, called, appropriately, bittern. Contained in bittern is magnesium, potassium, sulphates, boron, strontium, bromine and iodine. While, at the moment, John Ross' company doesn't use it, CSIRO scientists Hal Aral and Graham Sparrow believe bittern is a prime example of how the effects of salinity can be exploited commercially. Dr Aral believes the substances dissolved in saline groundwater can be used for fertilisers, light metals, plastics, industrial chemicals, oil refining, pesticides, glass, ceramics, bleach, soap, dyes, sewage treatment, sugar refining and alcohol brewing. 'The list is almost endless,' he states. Dr Aral dreams of one day seeing networks of evaporation ponds, solar-powered desalination plants and energy-storage ponds across the Murray–Darling Basin to provide fresh water and valuable chemicals.

Duncan Thomson is one man trying to turn salinity into a commercial reality. His company SunSalt has been making salt for a decade near Hattah in northwestern Victoria. The process is the same as at Pyramid Hill with saline groundwater pumped to the surface and poured into evaporation ponds. The salt is harvested and sold for table and industrial use. Although his business is successful, Duncan accepts that he struggles for market share against the large seasalt companies. Transport is the biggest enemy with costs of up to $55 per tonne to move the salt to markets.

Duncan has a plan though, or in his words, 'another arrow to shoot'. He has developed Australia's first transportable magnesium-harvesting plant: 'We built it in Mildura, put it on the back of a truck and brought it down here to Hattah, lifted it off, joined the bits and we are ready to go.'

After the groundwater has been brought into the evaporation ponds and the salt harvested, the remaining bittern is poured into another pond for further evaporation until it is almost dry. What remains is then initially processed in the transportable plant before being trucked back to a secondary processing plant at Mildura which produces crystals ready for sale. One product it can be used for is magnesium sulphate, which is extensively used in horticulture particularly as a fertiliser for grapes and citrus. Australia currently imports 3000 tonnes per year of magnesium sulphate. SunSalt's plant also produces magnesium chloride that can be used as a dust suppressant for dirt roads because of its ability to stay moist. 'People don't believe you are making fertiliser out of salt water but when you look at potash it is potassium carbonate. We bring 340 000 tonnes of the stuff in from overseas,' says Duncan. 'You can look at it this way, we are trying to make people believe that salt is not a scourge and that we can look at it another way.'

Graham Sparrow from CSIRO says the market would probably be saturated with magnesium products if there were two or three operations going on in the Murray–Darling Basin: 'That's why we need to explore all options. We need to see a number of plants along the Murray, all mining a range of chemicals and using salt along the way. There is so much salt that would be a by-product of the mining that we need to find a way to contain it in some form so it doesn't go back into the system. These are all words and ideas, what we need to do is some calculations and tests and find these higher value products.'

Across the river about 30 kilometres east of Pooncarie in southwest New South Wales, a joint-venture company called Murray Basin Titanium has begun mining and processing mineral sands. Rutile, zircon and ilmenite are mined from three deposits covering roughly 160 square kilometres and then taken away for processing. During that processing, saline water is used. According to the Australian Institute of Geoscientists, the deposits of mineral sands in the Murray–Darling Basin are fossilised beaches from a 6 million-year-old inland sea called the Moravian Gulf. If this inland sea still existed it would spread from Adelaide to Broken Hill. When the seawater disappeared it left behind an estimated 45 million tonnes of minerals including a black compound of titanium, iron and oxygen called ilmenite. According to Chief of CSIRO Minerals Rod Hill, herein lies the key to turning very old sand into very large amounts of money. Titanium is a strong, light and corrosion-resistant metal that is rare and sought after, and naturally expensive. Dr Hill believes the Holy Grail is to find a way to replace the current complicated way of making titanium with a single process directly from ilmenite.

It's blue-sky stuff and maybe the economics of such plans appear overwhelming. However weighed against the destructive cost of salinity is Dr Hill's estimate that mineral sands deposits in the Murray–Darling Basin are worth $13 billion. Dr Hill acknowledges processing bittern or mineral sands will not, in themselves, solve the salinity problem of the Murray–Darling Basin as a whole, but he does believe they could ameliorate the problem locally around the processing plants and offer a financial alternative to unsustainable land uses.

CASE STUDY: PAKENHAM, VICTORIA

RECLAIMING THE GREENS—NATURALLY

The Packenham and District Golf Club southeast of Melbourne looks a picture. The greens and fairways are well covered with grass, the trees are growing well and the golfers are out in force. However the course today looks a lot different than it did 10 years ago, as Course Superintendent Anthony Wright explains: 'When the Pakenham golf course was established in 1985, it was built on a floodplain covered in tea-trees (*Melaleuca ericafolia*). A

Pakenham Golf Club superintendent Anthony Wright and Kirsten Barker, salinity officer (Department of Natural Resources and Environment) for the Port Phillip region, look over one of the club's greens.

lot of these trees were removed to build the golf course, but without a plan for managing the effect on the watertable, the effects of salinity started to show up very early on. We have tackled the problem by improving the soil balance to deal with the high sodium levels in the soil. Our approach has been to address the problem, not to apply a bandaid.

'By getting a good balance in the soil we have been able to show a big improvement in the cover on the fairways. With better soils and good grass growth we have also been able to utilise effluent water in our irrigation program and that wouldn't have been possible a few years ago.

'The inspiration for our approach has been William Albrecht, an American soil scientist and work he did in the 1940s and 1950s. Albrecht had taken soil samples around the world and determined there was a common thread for fertile soils in terms of the levels of calcium and magnesium in the soil.

'We started with a soil that had a pH of 2 with high levels of sodium and sulphur. In earlier days gypsum was being added but that was making the problem worse. By meeting Albrecht's balance for calcium and potassium, the pH has improved and we have been able to grow good grass on the fairways and greens.

'Some of the trees on the golf course were suffering from the high salt levels, so we went back to planting *Melaleuca ericafolia* and swamp gums (*Eucalyptus ovata*) and they are doing much better than trees from outside the area. We have planted about 4000 trees so far.

'We use mostly organic fertilisers and haven't used insecticides or chemical fertilisers for two years. I would like to have the first certified completely organic golf course in Australia, but so far we haven't found a way to control weeds such as paspalum without herbicides. Our main fertiliser is chicken litter from an organic poultry shed nearby. We spread it annually on the fairways. Soil testing is undertaken once a year on the fairways and quarterly on the greens which allows us to keep a pretty close watch on fertility levels and tells us when some intensive treatment is required.

'We still have a drainage problem and the golfers will tell you it gets pretty wet during winter. But we can't flush or pump out

the excess water because there is nowhere for it to go. A quote to reshape and drain one fairway is around $75 000 and that's out of our reach. You can't beat good drainage, but by tackling the soil chemistry we can make a difference. We also match our irrigation with evaporation and tend to err on the side of not enough water. Kirsten Barker our local salinity officer from the Department of Natural Resources and Environment is keeping an eye on watertable levels. She reports regularly on levels in monitoring bores on the golf course and in the nearby residential area.

'We have noticed a fall in the watertable, but with several dry years it is hard to accurately pin down the reasons. What we do know is that about 2000 hectares of discharge have been mapped in the Western Port catchment and there is sure to be more than what has already been mapped.

'It really hasn't been that difficult or expensive and the proof is in the pudding. We have a better gold course for our members and in 1999 the Cardinia Shire recognised our efforts with an Environmental Management Award. ■

The state of the course before (opposite left) and after (above) the club's extensive restoration work.

BENEFITING FROM DIRTY NEIGHBOURS

While scientists dream of what might be, there are those who look at what already is and ask: 'How can we make a buck out of this?' It isn't immediately obvious how you can make money out of global warming, or how in doing so, you can help reduce salinity but there is a way.

In 1997 the nations of the world gathered in the Japanese city of Kyoto to talk about the weather, specifically just how hot it's been lately. Most scientists now agree that 200 years of industrialisation and large-scale land clearing has altered the planet's climate. Burning fossil fuels and pumping gases such as carbon dioxide into the atmosphere has wrapped a sort of atmospheric blanket around the globe. The result is that the hottest 10 years since measurements have been taken have all been recorded since 1980. At Kyoto, governments pledged to reduce emissions of greenhouse gases and look for ways to create green fuel sources. Australia hasn't gained any environmental glory by arguing for higher allowances than other countries, and then refusing to sign the protocol because, according to Prime Minister John Howard, it would be bad for business.

This has not proven to be the case for some Australian businesses. Part of the plan to reduce emissions is to give them currency by turning them into a commodity that can be bought and sold on the open market like stocks and shares. These are called carbon credits. If a country is overstepping the mark and producing too much carbon dioxide, then they would be in carbon debt. Like a bank account that is overdrawn, they would be on the lookout for some carbon credit to balance the books. The best forms of carbon credit are trees because they lock up carbon from the atmosphere. In this global approach, a tree anywhere on earth is helping.

The Kyoto plan means you can make a buck out of your dirty neighbours and help reduce salinity at the same time. Polluters all over the world are looking for large, sparsely populated places where people will plant trees on their behalf.

'We are here to help,' laughs John Bartle from the Western Australian Department of Conservation and Land Management.

'Carbon credits fit in very nicely with the really strong pressures that already exist to get more trees back into agricultural landscapes for salinity control and better protection of landscapes.'

In an irony that would have pioneer Western Australian farmers turning in their graves, it seems the mallee is the best tree available for storing carbon. The tree that caused so much heartache to remove to create the wheatbelt is now on its way back in. That damned backbreaking bulbous root system stores the carbon while the trunk system above ground can be harvested for other purposes. Mallee has already been widely planted on cereal farms across Western Australia to lower watertables and reduce salinity. Now it could be another revenue stream by being sold for carbon credits to a polluter on the other side of the world. Western Australian wheat farmer Ian Stanley has planted tens of thousands of mallee trees believing they are a long-term prospect: 'Even now no one would be able to tell you whether you will make much money out of planting trees. It would be nice to think we'd make enough out of them to cover the cost of putting them in and maybe eventually the opportunity cost of the land they occupy that would otherwise be growing a crop.'

In June 1998, just six months after the Kyoto convention, Sydney energy company Pacific Power made Australia's first move on a 'Kyoto forest' by buying the carbon rights to 1000 hectares of eucalypt plantation on the New South Wales north coast. Since then, large multinationals such as BHP, BP and Toyota have also invested in carbon credits. BP says it wants to plant somewhere between 25 000 and 60 000 hectares in Western Australia; Toyota says it will spend $23 million over a decade on plantations worldwide.

The extreme views on the viability of carbon credits range from those who believe it could be a multi-billion dollar international industry to those who say it is isn't worth the paper it is written on. Hamish Cresswell from CSIRO confirms there are lots of companies watching and positioning themselves for future markets and admits while there is very little trading going on yet 'it is something we expect to happen in the future and it provides virtually another income stream, another return from tree planting and tree growing, which can only be good.'

VIEWPOINT
Dr Syd Shea

SYD SHEA WAS EXECUTIVE DIRECTOR OF THE WESTERN
AUSTRALIAN DEPARTMENT OF CONSERVATION AND LAND
MANAGEMENT AND IS NOW PROFESSOR OF ENVIRONMENTAL
MANAGEMENT AT THE UNIVERSITY OF NOTRE DAME AUSTRALIA.
HE IS CHAIRMAN OF THE OIL MALLEE COMPANY, WHICH HAS A
PROJECT TO PLANT 500 MILLION MALLEE TREES IN THE
WESTERN AUSTRALIAN WHEATBELT BY 2025.

I think we have to reinvent agriculture and I am optimistic about that. I certainly don't believe we should knock what has been achieved. This was probably the most inhospitable place on earth to have agriculture and now farmers and our agricultural scientists have made our agricultural industry one of the most efficient systems in the world. However we can't go on the way we are going, both for environmental and economic reasons.

What we are proposing with the mallee project is to put in a new crop and a new industry that can be integrated with the existing agricultural systems and do a number of things. We can attack land degradation, make a significant contribution to reducing salinity and create a profitable crop because, frankly, we are not going to solve salinity unless the solution is profitable.

What we are doing is putting a natural species back into the ecosystem to consume a lot of water that is currently causing the problem. The beauty of mallee is it survives in one of the toughest environments in the world but also reshoots after it is harvested and it will reproduce time and time again. We are looking at producing renewable energy, activated carbon, and eucalyptus oil and of course the new product on the market is carbon credits. The mallee tree is designed to create carbon sinks. Essentially what we are talking about in terms of carbon sinks is trees or any perennial crop sucking carbon out of the atmosphere and storing it. It just so happens that mallee is a brilliant tree for this because you can actually have two sinks. That tremendous underground root system that caused my grandfather so much trouble, which lasts for hundreds of years, is one sink and there's another sink above ground that can then grow or you can harvest.

I just can't emphasise enough that the only way to solve the salinity problem is to make the solution profitable. There's not enough money in government, there's certainly not enough money in the farming community, nor can we expect them to contribute, but if we can make it profitable there are lots of investors around the world looking for good ways to invest their money and that is what we are targeting.

The mallee's a unique tree that can create a range of products, a new income for farmers, contribute to the reduction of greenhouse gases and salinity, and at the same time restore biodiversity, so forgive me for being enthusiastic.

Mallee is grown for more than just carbon credits. The tree is harvested by being sliced off at ground level, and the best time to do this is while the tree is only a few years old so harvesting is regular. While work is going on above ground, below the soil the giant root system is draining the landscape and reducing the threat of salinity. Scientists at the Western Australian Department of Conservation and Land Management estimate a mallee will survive several hundred years of such treatment. The leaves and branches can be converted into a number of products including eucalyptus oil, activated carbon (used in gold processing), charcoal and possibly steam generation for electricity. In February 2001 the CSIRO built a demonstration integrated wood processor (IWP) at Narrogin, nearly 200 kilometres southeast of Perth. An IWP uses tree biomass to produce electricity with activated carbon and eucalyptus oil as by-products. A full-scale IWP plant would use an estimated 100 000 tonnes of trees annually to produce 5 megawatts of power, 3500 tonnes of activated carbon and 1000 tonnes of eucalyptus oil. Mallee is ideal fuel for such a plant. The biomass of eucalypts can be converted into methanol (wood alcohol) or ethanol (ethyl alcohol), which is used as a fuel additive and may become more viable as fossil fuel deposits begin to shrink.

Mallee isn't the only tree-planting industry that could capitalise on carbon credit trading and return a second income. While the spicy, sweet smell of sandalwood burning is mostly associated with eastern religious worship, it's also used for oils, perfumes, soaps and woodcarving. The largest user is India but growing demand in China means the world market is increasing by 5 per cent per year. Although it seems an exotic tree, there are actually six Australian varieties of sandalwood. Being parasitic the tree attaches itself to the root system of another tree, and in Australia wattle is a favourite. Sandalwood takes its time growing and isn't ready for harvesting for almost 20 years but it's worth the wait given that good quality logs currently sell for $10 000 per tonne. Tea-tree, though not as lucrative as sandalwood, is another developing tree-planting industry and is far quicker at returning a profit. The small paperbark tree produces oil in its leaves that can be distilled and used for antiseptic and medicated soaps.

According to Syd Shea from the University of Notre Dame Australia in Western Australia, carbon credits could be the circuit breaker that allows widespread tree plantations to become viable in Australia: 'We are talking about needing to plant three million hectares of trees and that's not going to be done just with government money—there just isn't enough money in the farming community to plant on that scale. We have to get investment but there's a Catch-22 situation where you can't get the trees in the ground for processing and you can't get processing until you have trees in the ground. The beauty of carbon credits is the product doesn't have to be processed and it doesn't have to be transported. It's formed every nanosecond and so there is the opportunity. Once we've got the tree in the ground we have broken the circuit.'

John Williams from CSIRO Land and Water agrees with Syd Shea that those on the front line will adapt and take up anti-salinity schemes initially to protect their investments, but will become far more enthusiastic if there is a buck in it: 'Until we have land uses that make money and address the cause of the problem we don't really have enduring solutions. We need a concerted research effort at rebuilding land uses for the Australian landscape that captures that water and turns it into wealth. That's where I think we need one very serious focus of attention.'

CASE STUDY: KUNMALLUP, WESTERN AUSTRALIA

RAISED BEDS BRING SWEET DREAMS

Russel and Margaret Thomson graze sheep and grow mixed crops on their 5500 hectare property, Kunmallup, 230 kilometres southeast of Perth. Raised beds in dryland cropping have proved a way of overcoming waterlogging problems as Russel explains: 'A few years ago, if anyone had mentioned raised beds, I would have thought they were talking about fashions in furniture! But now I see raised beds as one of the best ideas for turning waterlogged country into good cropping land.

'Back in 1997 we first experimented with a 10 hectare site that had experienced frequent crop failures due to waterlogging. In the first year on raised beds our oat yields were 47 per cent higher. In the second year we had establishment problems with canola and had to reseed. The beds yielded about 40 per cent higher that year and results were similar from field peas the next year. By 1999 we had enough confidence to go into 30 hectares of salty country. I built a bed-former for about $13 000 by adapting existing machinery. It was harder than we expected—trying to form beds around trees just didn't work! A drain crossing the paddock on an angle to the beds didn't help either! We've learned a lot the hard way.

'My family has owned this property since 1955 and about one-third has been affected by salt. In the late 1950s my father was cutting drains to remove the water, and in the early 1970s I bought a bigger and better grader. Over the years we have also fenced off scalded areas and planted puccinellia and tagasaste to use water. Landcare work is self-sustaining if there is a dollar in it. Planting trees is nice but the investment needs to be effective.

Russel Thomson compares the barley seedling growth between raised beds and normal cultivation.

View of vigorous barley growth on raised beds on a previously waterlogged and saline paddock.

'Until now I have been only able to crop 40 to 50 per cent of the property because of waterlogging on the flats. The lower land gets run-on water that has nowhere to go. Cropping is hopeless unless you can get the water off the surface and out of the root zone.

'One of the main challenges with raised beds is seeding depth in the soft soil when equipment has been designed for hard seedbeds.

'This year we've put in a further 100 hectares of raised beds on salty country and are comparing a couple of different bed designs plus pasture growth. It's very flat country but raised beds are turning very poor grazing country into land we can crop. It's crucial to get the drains right and we've used a high-precision contour survey. I believe we can turn around these salty clay areas but it will take more than a year.

'The principle of beds is a good one. I think of it as "irrigation backwards". If you can get a 50 per cent increase in productivity you'll pay for new gear pretty quickly. But you have to put the time in and pick the right sites.' ■

CHAPTER 8
The Future

*Human beings, who are almost unique in
having the ability to learn from the
experience of others, are also remarkable
for their apparent disinclination to do so.*

DOUGLAS ADAMS

Trying to find elaborate processes to get trees planted is unnecessary if you don't cut them down in the first place. It is easy to look at old black-and-white film of tractors grunting through the bush dragging balls and chains behind them and dismiss it as ignorant folly. It isn't as easy to look at the same thing in digital colour knowing that Australia still clears more land than it reforests. Although state governments continue passing legislation to protect native vegetation to various extents, there are those who remain unconvinced. Australian Conservation Foundation President Peter Garrett says for every tree planted in Australia at least 100 are bulldozed and argues that Australia clears more forest than any other developed country. The chief culprit is Queensland, where far more bush is cleared than any other state or territory. At its height in 1999–2000 more than 350 000 hectares of bush was levelled according to the Statewide Landcover and Trees Study carried out by the Queensland Department of Natural Resources and Mines.

This accounted for about 60 per cent of the national total that the Australian Conservation Foundation puts at 564 800 hectares (sixth highest in the world). Queensland has often been a stand-alone state and thrived on being different from the rest of the nation. So it is—or rather was—with salinity.

The landmark Australian Dryland Salinity Assessment published in 2000 (the same year as the record-breaking land clearing), confirmed that Queensland was a far less salty place than elsewhere. It showed nowhere near the cancerous marks on the map that are so upsetting for its southern neighbours. It is also true that the base data to create such a map was difficult to find in Queensland. The best informed guesses of the national audit showed that the state had about 3 million hectares that are at high risk of developing salinity by 2050. These areas are mostly in agricultural land scattered from Brisbane to Cairns. The Condamine River, Lockyer Creek, lower Mary River, South Burnett River, Three Moon Creek and some tributaries of the Fitzroy River showed an increase in salinity. There had been no state costing on salinity impacts. The audit did refer to an 'urgent need to establish a state-wide network monitoring groundwater, surface water, key land use and biodiversity parameters to better inform managers of the trends and implications of dryland salinity'. Its next recommendation was that 'preventative and protective action is taken in maintaining water balance for those areas where salt stores would be mobilised with the clearing of native vegetation'.

John Williams from CSIRO Land and Water is less subtle when asked about Queensland's potential for salinity: 'Stop clearing,' he almost shrieks. 'Having worked in salinity for almost 25 years, what I have seen is once it starts it's very, very difficult to halt. Now in Queensland there are many areas where we have the potential to prevent it ever happening and the key to preventing it ever happening is to stop clearing. We have been clearing parts of Queensland at something like 300 000 hectares a year, which to me is very dangerous because I believe the salinity hazard in Queensland is very real. We have an opportunity to stop the clearing, do the analysis and prevent salinisation over large areas. The problem facing Queensland and the people working there, and

I have great sympathy for them, is that it is different from southern Australia. In southern Australia, people can see the symptoms. What we have got to do in Queensland is help them understand the landscape and scientific methodology and teach the people about the hazard of salinity because once you see the symptoms of salinity in Queensland it is too late. It is just like the rest of Australia, so prevention in northern Australia is to me a very, very high priority.'

In July 2002 Queensland Premier Peter Beattie rolled out a map of the Murray–Darling Basin showing areas at risk of salinising. 'This', he intoned to the waiting media, 'kills the myth forever that Queensland will be spared from the scourge of salinity.' The map was based on new research and confirmed the fears of John Williams and others that salinity was never going to stop at the New South Wales border. Roughly 80 per cent of the Murray–Darling catchment had some level of salinity risk. Peter Beattie interprets the information to mean that without major changes to land and water use, vast areas of Queensland's southwest would be useless by the middle of the twenty-first century. He said the hazard map suggested 'salinity could turn up to 26 million hectares of Queensland's section of the [Murray–Darling] basin into a wasteland' and concluded, 'We must act now to minimise and prevent salinity'.

THE TOP END

With Queensland now joining the southern states in feeling the nasty bite of salinity, it seems only the Top End remains unscathed. The National Land and Water Resources Audit showed that in 2000 the Northern Territory didn't have a single area that could be classified as a high-salinity hazard. The area that could be classified as having a moderate hazard was only 6 per cent and is centred on the Sturt Plateau near Daly Waters. The rest of the territory was divided into low hazard (34 per cent) and very low (60 per cent). Scientists believe even if large areas were cleared for agriculture it is unlikely to be a problem because the vegetation that grows so well in the high rainfall areas absorbs most of the water. Even the more arid regions show no sign of a rising watertable.

LOWERING THE FLAG

John Lawrie had everything at stake when salinity started attacking his rice farm at Coleambally, south of the Murrumbidgee River in the New South Wales' Riverina district. It isn't an overstatement to call John a giant of the rice industry because he is. His imposing 7 foot frame is such a striking physical specimen that in his mid-teens he went to the city to train with the Melbourne Football Club. They tried to beef him up and when he was not at footy training or doing schoolwork he also played basketball. 'I was so tired I fell asleep at my desk,' he remembers. John gave away footy and went farming. He notes with some sadness that of his graduating class in agricultural science from university, only two became farmers. The men and women who took up land when the irrigation settlement of Coleambally was established in the 1960s didn't all arrive with the qualifications of John Lawrie. Some were taxi drivers from Sydney, others were small businessmen, but they all shared in the sort of vision the Chaffeys could never have imagined almost a century before. At the time, the Snowy Mountains Scheme was nearing completion and with it the grand plan of drought-proofing inland Australia. John Lawrie says he was a gung-ho irrigator and with the promise of endless water he started growing rice. It was a success until, he says, 'salinity came up and bit me in the bum'. During winter the rice farm dries out and the drains at the edges of the fields have a white crust of salt on the surface. John Lawrie shows visitors a piezometer that lies between a drain and the dry paddy field. He bends down and grabs the flag that is fluttering at about ankle height and pulls the pipe out of the ground until the flag is above his head. Given his height this is a dramatic demonstration. He explains: 'As a result of our continuing irrigation we could see how our watertable had risen to the stage where it was only 20 centimetres below the surface and in doing that we could then see that this flag was way up here. We thought we were going to drown, let alone all of the other consequences associated with a high watertable such as increased salinity and all the rest of it.' The rest of it included native grasses dying, crop yields deteriorating and soil structure changing. After a pause for effect, John slowly lowers the pipe back down until the flag again sits just above the earth, then continues. 'So we had to introduce an initiative on the place to try and get that flag back down in the hole where you see it today. And that's happened over a period of six to eight months.'

The irrigators of Coleambally created their own land and water management plan that was based on saving their farms from salinity. It involved a whole-farm plan that uses several common tools such as crop rotation to absorb extra water during the off-season and exact watering during the irrigation season. They educate, monitor, research, develop, recycle, drain and upgrade—anything that will help. The irrigators say it improved water efficiency by 60 per cent between 1996 and 2001. As the watertable drops, smaller amounts of water are used to push the salt down below the root zone. It is estimated Coleambally imports about 100 000 tonnes of salt per year with its irrigation water from the Murrumbidgee. Only about 3000 tonnes of that returns to the river—the rest is sunk into the groundwater. The load builds every season and so the balancing act is a delicate one. It's the reason why flags on top of piezometers flutter across the landscape giving an immediate indication of how things are going.

It is easy to explain away the good work as enlightened self-interest but John Lawrie has become a born-again irrigator. During his annual holidays he often takes his family to Adelaide and shows his children who else relies on river water for survival. John and his industry colleagues are well aware of the invisible nature of Australia's silent flood: 'People now realise there is an interdependence on one another whether you live in the cities, whether you live in the country or whether you live somewhere in between. We all influence one another's livelihoods.

'We have adopted the attitude that we will not let surface water off this property whereas in previous years, prior to 1990, 30 per cent of the water used to go off this place. Whether it was full of salt or chemicals or whatever, I would have said to you, so what? But there is a whole change in the area. There is a change in my attitude. We all have to peacefully co-exist on this planet and we all have to produce food.'

VIEWPOINT
Rick Farley

RICK FARLEY IS THE FORMER EXECUTIVE DIRECTOR OF THE NATIONAL
FARMERS FEDERATION AND CATTLEMEN'S UNION OF AUSTRALIA.
ALONG WITH PHILIP TOYNE, FORMER DIRECTOR OF THE AUSTRALIAN
CONSERVATION FOUNDATION, HE CREATED THE LAND MANAGEMENT
PROGRAM, LANDCARE, IN 1988. HE IS NOW A PRIVATE CONSULTANT.

> *I think we have got a fairly good idea about the
> scale of the problem but we are not doing a very
> good job of combatting salinity. We've got a situation where
> land management is a constitutional responsibility of the states
> and it's very difficult to get a common approach from the states
> and the Commonwealth. In Queensland there is still large-
> scale land clearing going on, despite the evidence that it can
> only increase salinity. So I think the challenge for Australia is
> to take a much more integrated approach to salinity. I think a
> lot of the science is there—we know broadly the scale of the
> problem and we know broadly what we should be doing—it's
> just a question of getting out there and doing it now.*
>
> *There are two essential building blocks: first of all, a
> long-term policy framework involving the states,*

Commonwealth and local government because local government planning such as sewerage, all comes into it. We've got to involve the financial institutions, environmentalists, private landholders and Aboriginal people who are significant landholders. Then we need to have long-term funding commitments because every farmer will say it's hard to be green when you are in the red. There are a lot of farmers who want to change their management systems but haven't got the capital to do that.

When Philip Toyne and I talked to Bob Hawke about starting a national land care program we estimated it would cost $340 million over 10 years and in 1989, $340 million sounded like a lot of money but that hasn't even scratched the surface. Now to me, if you need to have a secure funding base, the best way to approach it is with a tax levy—that way the debate over money for salinity is taken out of the political debate about budgets each year. Taking that a step further, it implies there's got to be a consensus between all of the major political parties that this is what is required. I am a very strong advocate of a tax levy provided that levy is transparent and regular reports are provided to the electorate on how the money is being spent and the results being achieved.

I always recall a conversation I had with a very senior treasury official after the first GST debate with John Hewson in 1993. Sitting around talking about it afterwards, he said something I have never forgotten: 'In Australia we eventually do the right thing because it's the only thing left to do.' To me the heart of this issue is how long it takes us to do what we need to do. We will do the right thing eventually but it's a question of how much country is left.

CASE STUDY: HUNTER VALLEY, NEW SOUTH WALES

BREAKING ALL THE RULES

John Williams from CSIRO is the first to admit that scientists don't have all the answers to salinity. 'In this whole search for solutions it's terribly important that we gather them from wherever they exist,' he says. 'Some of them exist in the scientific community but many don't. Many are out there with people who are able to read the landscape and understand how it works.'

One such man is horse-breeder Peter Andrews. He operates a stud at Tarwyn Park near Bylong in the Hunter Valley in New South Wales. During its long history, the 250-hectare property has been at various times the home to four Melbourne Cup winners. In years gone by the thoroughbreds could have feasted on lucerne but by the time Peter arrived in 1973, salt was destroying the best land: 'There were 120 acres called Red Hill that hadn't grown green feed in 30 years, and there had been a change in the draining of a marsh area that really only started in the 1950s and 1960s, and then they had severe salinity by the 1970s.'

Peter has left a small section of dryland salinity on the property to remind himself and to show visitors just how badly things were going. The salt crusts are in sharp contrast to the lush paddocks nearby. The extraordinary thing is that Peter restored Tarwyn Park by ignoring the basics of treating dryland salinity— in fact he up-ends some of them. Instead of lowering the watertable he raised it.

Peter believes his best chance of restoring the landscape is to mimic the way water originally travelled through this land before farming had changed the drainage system. 'This country had a [natural] sustainable irrigation system and I did my best to identify it and reintroduce it,' he explains.

As he wandered the property working his horses Peter noticed dry areas on the high ground. When he looked closer he found that there was sand just underneath the topsoil and if he poured water on it, it very quickly disappeared. He reasoned that by adding water at these points he would fill his watertable with fresh water. Basic science told him that salty water is heavier than fresh water and so as the watertable rose, it had a freshwater layer on top near the plant root zone. Peter says by reverse leaching he has effectively created grass-covered dams. The next move was to redirect the stormwater towards the intake points and slow it down enough so it continued to fill the watertable. Small channels were dug to divert water and plants were added around the sandy patches so the water would sit there long enough to be absorbed. To avoid evaporation Peter adopted the philosophy that if it's green and grows then it is good. So long as the pastures were covered, the evaporation was minimal.

The results have been spectacular and Peter says the health of his horses has improved along with their surroundings. ■

Peter Andrews has found unorthodox methods to be most successful in combatting the dryland salinity that was ruining his horse stud.

IS THERE ENOUGH SCIENCE?

Matt Linnegar from the Ricegrowers' Association of Australia believes farmers are feeling more positive about tackling salinity because of the advances in science. However despite the strides made, he believes that more science is needed to create more effective long-term strategies: 'Before rice growers like John Lawrie put their faith in a management plan they want to know it is well grounded.'

The question of science sometimes elicits awkward answers from those involved with managing salinity. There is a belief among some that a salinity industry has been created in Australia, while endless research projects yield little new information and then require further long-term research. Naturally the country's senior scientists reject this. They fear tens of millions of dollars could be wasted in blindly attacking the salt without a scientifically based plan. Water ecologist Peter Cullen says some of the technical answers have been given but argues there is a need for more comprehensive farming systems that will use up all the water that falls or that is used for irrigation: 'There is no doubt we need more knowledge but we need different types of knowledge at the scales of whole catchments rather than small paddocks. We need to find better ways to target our investment on salt. We need better mechanisms for measuring success. How do we know if some of these things have worked over a 20- or 30-year time scale? We haven't had that long-term monitoring and whole-catchment instrumentation that allows us to work on the scale that we need to address these problems.

'I think it's very easy when looking at salinity in this country to put energy into attacking the symptoms. We need to put a lot more energy into attacking the causes. We've got a lot of engineering works in the lower areas where we have a lot of evaporation basins and drainage systems and we get proposals for more exotic things like pipelines to the sea and whatever. That's learning to live with the problem and that's always going to be part of the solution. But it seems to me, we are putting bandages on what is a critical catchment problem and we need to find more effective farming systems and use the water that falls.'

CSIRO's John Williams agrees some questions have been answered but he is yet to see large-scale land use address the root of the problem: 'We need at least 30 per cent of the catchment to go into trees or related deep-rooted vegetation to substantially switch off the leakage. We have to have that sort of number and to have that in the most effective location in terms of hydrology is important, and to work that together with the community, who may have other aspirations for that piece of land.

'So the priority for me is to not only recognise the substantial nature of the land-use change but to get in place community processes that can actually work towards having real on-the-ground implementation. Until we can show that we can do it, it's harder and harder to marshal the resources to show that it is worthwhile, because so much of the investment to date has been fragmented and small scale, and we need to turn that around. We don't want to go for magic bullets. We need multiple outcomes in a catchment. Now the science of how to put that landscape back together, how to rebuild the landscape, is a very new science and very difficult science and one that needs a lot of research and development.'

THEY'VE BEEN ARGUING SINCE FEDERATION

When the Australian colonies decided to come under a federal roof in 1900, unfortunately they didn't bring the Murray River along with them. Up until then the river had been a border between nations with customs houses on either side collecting taxes. In hindsight it might have been better for everyone if the new Commonwealth government had taken control of the waterway but it didn't. Instead an awkward arrangement was made where New South Wales and Victoria shared the river and agreed to give South Australia an annual allowance. Queensland wasn't included even though it is the largest catchment area of the Murray–Darling Basin. Sectional interests have never been overcome and have remained a stumbling block for widespread salinity control. Regional needs are fiercely attacked and defended and fingers of blame are pointed at almost every jurisdiction. The body created to manage the often-conflicting desires of four states, the ACT and the Commonwealth government is the Murray–Darling Basin Commission. Its chief executive, Don Blackmore, has heard almost every argument from every lobbyist and smiles when asked about sectional interests: 'Most people's interest in catchment management is normally judged by them standing in a catchment and looking upstream. That is what drives self-interest and it's true around the world. What the commission has been able to do is look 180 degrees at the sense of need and responsibility.'

Salinity is only one of the many issues the commission has to try to manage and so often policy decisions are made because they will deliver multiple outcomes. There have been three major developments that take on the sort of big picture needs that Peter Cullen and John Williams are concerned about. They also back up Matt Linnegar's theory that, in the longer term, groups will start to work together on single issues. 'In the shorter term I don't think the finger pointing exercise is quite over. So long as we have those sectional interests continuing to blame someone else for the issue that's going to be a problem. However I think it is starting to turn and we are starting to see groups working together.' Australia may even start to resemble a nation.

In 1995 the Ministerial Council, which comprises representatives of all governments and controls the Murray–Darling Basin Commission, agreed to cap the amount of water being taken from rivers in the basin. Don Blackmore believes the cap is the most important environmental decision ever taken in the basin. Since the cap, new players have had to buy their water from others and it has allowed many inefficient or saline-damaged irrigators to stop farming because they can still make money from their water allowance without increasing salinity. Every year in Australia, the equivalent of more than two Sydney Harbours full of water is traded.

'It hasn't solved the salinity problem and it hasn't solved the environmental problem with our river,' says Don Blackmore, 'but it has at least given us a point of stability in stream flows.' It's hoped that the demands of market forces for food and fibre set against a stable water amount will force greater water efficiencies on users in the basin, which in turn wil help slow salinity.

Having stabilised the amount being taken, how then can more water be given back to the rivers for the sake of their health? The new flow regimes created after the 1920s when artificial barriers started being built in rivers across the basin has taken the highs and lows out of the boom–bust cycle. The modest flows have allowed salt levels to build, and there are few floods of a size that can both dilute and flush the system.

By the beginning of the twenty-first century it was clear that more water was desperately needed down the system. In spring 2002 as the country struggled through a winter of drought, the Murray Mouth almost gasped its last breath. No river water had flown past the Goolwa barrages since the previous Christmas and the Southern Ocean had pushed so much sand through the mouth, it had all but closed shut. Any salt upstream was staying there. The Coorong was in danger of cooking over summer with high water temperatures and rising salinisation from evaporation. In a desperate measure, $2 million was spent dredging a channel through the sand to allow fresh ocean water to flow into the Coorong. It was a stark and very public reminder that the river was in serious trouble.

Several weeks before the crisis, in a Goolwa watering hole, a public meeting of some significance was held to try to find out how to get more water into the river. Local irrigators, environmentalists, commercial fishermen and others concerned about their river listened to, and in turn lectured, experts from the Murray–Darling Basin Commission. This was the first of what would be hundreds of meetings to see how, for the first time, the cap could be lowered. The theme of the evening was on creating both a healthy and a working river, suggesting that all parties would be listened to. The scientists said the best chance the Murray had of getting off the critical list was to have 1500 gigalitres of extra water flow—an equivalent of a 20 per cent cut in the cap. Those in the room knew there was an impossible amount of talking, and possibly shouting, to be done just to get a

VIEWPOINT
Corey Watts

COREY WATTS GREW UP IN THE SOUTHWEST OF WESTERN AUSTRALIA. AFTER STUDYING AT MURDOCH UNIVERSITY HE WORKED FOR WWF, THE AUSTRALIAN DEMOCRATS AND THE WESTERN AUSTRALIAN DEPARTMENT OF CONSERVATION AND LAND MANAGEMENT. HE IS THE CO-ORDINATOR OF THE AUSTRALIAN CONSERVATION FOUNDATION'S SALINITY AND SUSTAINABLE AGRICULTURE PROGRAMME.

Salinity is the end result of a major amount of environmental degradation. Around half of Australia's landscape has been cleared or significantly altered and it's the beginning of a new wave of environmental disruption.

Not only do we stand to lose literally thousands of native plant and animal species and all the values that people hold dear and associate with our beautiful landscape but also nature plays a key role in maintaining healthy rural communities and economies. If you remove native biodiversity you start to impact on Australian farming quite substantially.

CSIRO estimates that $1.5 billion a year is the contribution of environmental services like water quality and soil fertility to Australian agriculture.

We are going to have to live with a lot of damage and probably a lot more damage but we can minimise that damage if we do the right things, and the most essential thing is we have to undertake a major national land-use change. We have to develop farming systems and land-use systems that are suited to the Australian landscape rather than those that have been imported from elsewhere.

In the past we have focused on drainage and salt interception, engineering solutions if you like. They are really only effective in the short term and in some instances they have their own environmental problems as well. They do have a role but by and large what is required is a major revegetation effort across huge areas of our catchments. The need for that effort has not yet been fully grasped by our leaders and key decision-makers. It requires a major program of investment in new crops—perennial deep-rooted crops and trees—to produce both sustainability outcomes and commercial returns.

You can't have money without strategy. So we need a decent strategy but we also need a scale of investment unprecedented in this country in terms of environmental expenditure. That needs to come from both the public and private sector.

I think only recently have we begun to acknowledge that we are here to stay. We need to treat this country as though we are passing it on to the next generation and the one after that. We want them to have the same level of shared prosperity we have enjoyed, if not better.

recommendation to the Ministerial Council, who would then decide if it would go ahead. But it was a turning point because for the first time the governments that make up the council all agreed that the current system is unsustainable, that more water is needed and that they had to ask the stakeholders how to deliver it.

Whether the water savings are made by fixing leaking water delivery systems and infrastructure, or by reducing water allocations to current users has to be thrashed out. However it happens, the man in charge of Murray Water, David Dole, thinks it will rank with, or even higher than, the initial decision to impose a cap: 'It's an extraordinarily significant decision to rebalance the environmental and consumptive needs of the river. How long it will take us to get there and what we actually get to are questions for the future. I am sure it is going to need some profound leadership from governments.' Tim Fisher from the Australian Conservation Foundation agrees, adding it will 'be the biggest environmental rehabilitation project in Australia's history—and the most expensive. I guess at least $2 billion over 10 years will be needed'.

The third development is the other major decision of the Ministerial Council, made in March 2001, which was to create salinity targets for the Murray–Darling Basin's 21 catchments. If the cap was the biggest environmental decision, then Don Blackmore believes this was the most important single-policy response to salinity, but one that will cause pain. His eyes light up a little as he explains how it is applied in an area as big as France and Spain combined: 'It captures both the impact of point-source pollution and diffuse pollution at the end of every valley, and it says to the valley community you are responsible for living with that envelope of salinity—now it is your problem.

'The challenge is for that community to craft its solution. How many trees do I put in? What engineering solutions? What changed land practices? And so on. I think I am excited about the next decade because I think we are going to see those matters brought to account.'

The Murray–Darling Basin is going to have to live with a lot more salt. Too much is moving and too little can be done to stop it. Even land that has been destroyed by salt cannot just be abandoned. It needs to be stabilised rather like a toxic-waste dump so its salt load doesn't continue to spread. World history doesn't document stories about irrigation civilisations that are thousands of years old. To the contrary, those recorded lasted only a couple of hundred years before salt and silt destroyed them. A sobering thought as we pass the centenary of the Chaffey brothers' arrival in Mildura and Renmark. There are some haunting sites along the back roads of Australia's irrigation communities but it is hard to find anyone ignorant of the subject. In towns and on farms people are protecting themselves from it, school kids are doing projects on it, on weekends communities are planting trees, farmers are getting better at irrigation and scientists are producing useful information. Irrigators and environmentalists have at times been portrayed as rednecks versus greens but it is amazing how often they say the same things about salinity. Brad Williams from the NSW Irrigators' Council and Tim Fisher both believe the 1990s saw a dramatic change in attitude to salinity for the better. As Brad comments: 'Land and water management plans are having a positive effect on salinity levels in the [Murray] river but we need to keep working toward those outcomes. Things can and are happening and that's the most important thing.'

LOSING LAKE TAARBLIN AND SAVING LAKE TOOLIBIN

For a final word on Australia's salinity crisis it is necessary to head to the salt heartland—the Western Australian wheatbelt. Just past the town of Narrogin, nearly 200 kilometres southeast of Perth, is Lake Taarblin. There is almost no sound at this lake—no birds, no frogs and no insects. The dead trees are pale and bone dry and their fallen limbs crunch under your feet as you walk around them. The salty smell is distinctive.

Just 7 kilometres from Lake Taarblin is Lake Toolibin, where a battle is being waged to save the last freshwater lake in the region from salinity. The 300-hectare bowl is at the headwaters of the Blackwood River and its floor is covered with casuarinas and melaleucas. The lake is included in the intergovernmental Ramsar Convention's List of Wetlands of International Importance because of its significance as a place for migratory birds. However it lies at the bottom of a 50 000-hectare catchment where 95 per cent of the native vegetation has been cleared over the past century. Watertables are rising by 10 to 20 centimetres per year. Almost 10 per cent of the catchment is already severely affected by dryland salinity while another 24 per cent is at high risk, meaning the area is about one-third of the way through its salinisation process.

Without significant changes Toolibin could soon end up looking like Taarblin, so the local community has gathered professionals in to help them save the lake. They have taken a fortress approach with large surface-level engineering diverting early saline-runoff flows away from the lake and a groundwater pumping scheme is trying to lower the watertable.

The future of the lake doesn't depend on what has been done so far but what happens in the catchment. Farmer and industry groups are aiming for the land management changes that will protect the lake from the creeping salt. Oil mallee is being established, lucerne has been planted, perennial grasses are being used for grazing, farming is now done on the contour and a pine plantation has begun to halt recharge and control surface water. It is only the beginning and the coverage so far won't be enough to stop the salt. Scientists predict almost half the landscape needs to be covered to control the underground water movement. In the meantime the pumps are going flat out at a cost of $100 000 per year.

Standing in the centre of Lake Toolibin, Richard George from the Western Australia Department of Agriculture explains his admiration for the people who farm and live on the land surrounding the lake and who have made an enormous effort to save it: 'It's something that the community has decided is important. We are trying to make sure we hold the tide back as long as we can until we develop the economic systems that sit on the farms, that fit with the farming practices of this district and that eventually will prevent what is most certainly the death of this landscape if we get it wrong.'

He could be speaking for all Australia.

APPENDIX 1
Salinity tolerances for irrigated plants

Salinity Mg/litre	Crops and Pastures	Flowers and Shrubs	Fruit and Vegetables	Lawns and Grasses
300		violets	loquat	
700		aster, begonia, fuschia, rose, azalea, camellia, gladiolus, zinnia, bauhinia, dahlia, poinsettia	avocado, walnut, blackberry, strawberry, French beans, peas	
800				bent grass
850	field peas and beans			
1000	broad beans flax	bougainvillea, hibiscus, carnation, vinca, coprosma	apple, almond, apricot, grapefruit lemon, orange, peach, pear, plum, raspberry beans, capsicum, potato, celery radish, lettuce	
1200	clover			bluegrass, fescue, rye grass
1350	groundnut	chrysanthemum, oleander, stock	fig, grape, olive, pomegranate broccoli, onion, carrot, cauliflower, gherkin, sweet corn, cantaloupe, potato, cucumber	

Salinity Mg/litre	Crops and Pastures	Flowers and Shrubs	Fruit and Vegetables	Lawns and Grasses
1750	rice		artichoke, tomato	
2000	berseem clover, corn, millet, soy bean, sudax, lucerne, safflower			
2100			asparagus, cabbage, beetroot, spinach	
2800	phalaria, sorghum, sunflower			
3000				Tall fescue
3200	perennial rye grass, sudan grass			
3700	barley, cereals (wheat), cotton, sugar beet			
5000				Santa anna couch
25 000				Sun turf, kikuyu

Source: South Australian Department for Water Resources

APPENDIX 2
Salt-tolerant trees and shrubs

LARGE TREES—HEIGHT 15 METRES OR MORE

Botanical name	Common name	Natural occurrence
*Acacia auriculiformis***	northern black wattle, ear pod wattle	NQ, NT
*Casuarina cunninghamiana***	river sheoak	Q, NSW, NT
*Casuarina glauca***	swamp sheoak	SEQ, NSW, NT
Corymbia citriodora ssp *variegata*	spotted gum, lemon-scented gum	Q
Corymbia tessellaris	carbeen, Morton Bay ash	Q, NSW
Eucalyptus argophloia* `	western white gum	SQ
Eucalyptus brassiana	Cape York gum	NQ
Eucalyptus brockwayi	dundas mahogany	WA
*Eucalyptus camaldulensis***	river red gum	All mainland states
Eucalyptus cambageana	Coowarra box	CQ
*Eucalyptus drepanophylla***	Queensland grey ironbark	NQ, CQ, SEQ
Eucalyptus grandis	rose gum, flooded gum	NQ, CQ, SEQ
Eucalyptus largiflorens	black box	SQ, NSW, Vic., SA
Eucalyptus melliodora	yellow box, honey box	CQ, SEQ, NSW, Vic.
Eucalyptus microtheca	coolabah	Q NSW, NT, SA, WA
Eucalyptus moluccana*	grey box	NQ, CQ, SEQ, NSW
Eucalyptus paniculata	grey ironbark	NSW (coastal)
Eucalyptus pellita	red mahogany	NQ
Eucalyptus raveretiana	yellow box, black ironbark	NQ, CQ
Eucalyptus robusta	swamp mahogany	SEQ, NSW (coastal)
Eucalyptus salmonophloia	salmon gum	WA
Eucalyptus salubris	gimlet, fluted gum	WA
Eucalyptus sideroxylon	ironbark, mugga	SEQ, NSW, Vic
Eucalyptus tereticornis	forest red gum, blue gum	Q, NSW, VIC

Botanical name	Common name	Natural occurrence
*Melaleuca leucadendra***	broad-leaved tea-tree	NQ, CQ, NT, WA (tropics)
Melia azederach	white cedar	NQ, CQ, SEQ, NSW, WA

MEDIUM TREES—HEIGHT 5-15 METRES

Botanical name	Common name	Natural occurrence
*Acacia ampliceps***	salt wattle	WA, NT
Acacia disparrima sub sp *disparrima* (syn. *A. aulacocarpa)*	southern salwood	CQ, SEQ, NSW
Acacia crassicarpa	northern wattle	NQ
Acacia leptocarpa	wattle	NQ, CQ, SEQ, NT
Acacia pendula	weeping myall	CQ, SWQ, NSW, Vic.
Acacia salicina	cooba	All mainland states
*Acacia stenophylla**	river cooba	All mainland states (inland)
Callistemon salignus	white bottlebrush	SEQ, NSW
Callistemon viminalis	weeping bottlebrush	NQ, CQ, SEQ, NSW, Vic.
Carallia brachiata	carallia	NQ, NT, WA
Casuarina equisetifolia	beach sheoak	NQ, CQ, SEQ, NSW
Eucalyptus burdettiana	Burdett's gum	WA
Eucalyptus curtisii	plunkett mallee	SEQ
Eucalyptus sargentii	salt river gum	WA
Eucalyptus spathulata, ssp.*spathulata*	swamp mallee	WA
Melaleuca arcane	winti	NQ
Melaleuca bracteata	river tea-tree, white cloud tree	NQ, CQ, SQ, NSW, SA, WA
Melaleuca linariifolia	narrow-leaved tea-tree	NQ, CQ, SQ, NSW, NT
Melaleuca quinquenervia	broad-leaved tea-tree	NQ, CQ, SEQ, NSW
Pittosporum angustifolium	pittosporum, cattlebush, bitterbush	All mainland states (inland)

SMALL TREES AND SHRUBS—HEIGHT UP TO 5 METRES

Botanical name	Common name	Natural occurrence
Atriplex nummularia*	old-man saltbush	SQ, NSW, Vic., SA
Callistemon citrinus	lemon-scented bottlebrush	SEQ, NSW, Vic.
Callistemon phoeniceus	fiery bottlebrush	WA
Eucalyptus forrestiana	fuchsia mallee	WA
Leptospermum polygalifolium	tantoon, wild may	NQ, CQ, SEQ, NSW
Melaleuca nodosa**	prickly-leaved paperbark	CQ, SEQ, NSW

*above average growth and survival in saline areas

**best species for highly saline areas

Source: Queensland Department of Natural Resources and Mines, 2001.

Abbreviations: CQ = central Queensland; NQ = north Queensland; NSW = New South Wales; NT = Northern Territory; Q = Queensland; SA = South Australia; SEQ = southeast Queensland; SQ = southern Queensland (inc. SEQ); Vic = Victoria; WA = Western Australia.

GLOSSARY

ANAEROBIC Soil conditions where free oxygen is deficient and reducing conditions prevail. This occurs in waterlogged or poorly drained soils where water has replaced air in the soil pores.

ANOXIC An absence of oxygen in water.

AQUIFER A porous soil or rock formation that holds water and through which water can reach bores and springs.

CAPILLARY RISE The upward movement of water through a soil that is caused by molecular attraction between water and soil particles.

CATCHMENT An area of land supplying water to a watercourse bounded by hills or ridges that direct the flow of water.

CATCHMENT DISCHARGE POINT Point in the catchment where water is released.

CATCHMENT THROAT An area in the lowest portion of a catchment where water is released.

CONNATE SALTS Salts naturally present in the soil profile coming from marine sediments deposited in earlier geological times.

CYCLIC SALTS Salts transported from the ocean and deposited by rainfall.

DISCHARGE AREAS Area where groundwater 'discharges' to the land surface due to a restriction on 'down slope' water transmission. Discharge is maximised where the watertable is at or very close to the soil surface, where there is little vegetation, or where soil properties tend to encourage water to move towards the surface.

DISPERSION The process whereby soils separate into their constituent particles in water. Dispersible soils are highly erodible.

DRAINAGE BASIN A large area draining to the base of a catchment.

DRYLAND SALINITY Saline seepages or salt scalds occurring in rain-fed (non-irrigated) areas caused by changes in land use that affect the groundwater balance throughout the landscape. A typical situation occurs following the tree clearing from hillslopes, which reduces transpiration and allows an increase in rainfall intake beyond the root zone and a rise in watertables lower down the slope. Increased subsurface seepage dissolves salts in the soil and, with lateral flow through the landscape, moves from hillslopes to valley floors. Salty water then surfaces in patches depending on the geomorphology and topography of the site. The salt becomes concentrated by evaporation at these locations and the normal vegetation is killed.

EC Electrical conductivity. The most common measurement of salinity in soil and water. Salt is a good conductor of electricity, so the more salt, the higher the EC value.

EROSION The wearing away of the land by running water, rainfall, wind, ice or other geological agents.

EVAPOTRANSPIRATION Water returned to the atmosphere by evaporation (by the sun) and by plants emitting water vapour from their leaves.

FLOCCULATION The process by which very fine clay particles, suspended in water, loosely come together into larger masses which eventually sink down into the sediment.

GROUNDWATER Subsurface water in a saturated zone of the soil or rock.

HALOPHYTE A plant capable of living under salty conditions.

HALO-TOLERANT To be tolerant of saline conditions, that is, salt tolerant.

HYDRAULIC CONDUCTIVITY The rate at which a soil allows water to move through it.

HYPOXIA A state where oxygen is deficient.

INNATE SALTS Salts released during the process of soil and rock weathering.

IRRIGATION SALINITY A form of salinity that is caused by the increasing build-up of salts in soils used for irrigation. It results from raised watertable levels that bring soil salts into the upper levels of the soil profile, as well as the repeated use of saline river water for irrigation. This can commonly be the result of inefficient water-use practices.

LAND DEGRADATION A decline in the overall quality of soil, water or vegetation condition, commonly caused by human activities.

LANDCARE Landcare is a community-based approach to fixing environmental problems and protecting the future of our natural resources. There are now more than 4250 Landcare groups across Australia. About one in every three rural landholders is a member of a Landcare group.

LEACHING The removal of soluble minerals and salts by water seeping through a soil, rock, ore body or waste material.

PERCHED WATERTABLE A watertable which sits above (perched on top of) an impermeable rock or soil structure.

PIEZOMETER A non-pumping, deep bore (less than 3 metres) that is used to measure groundwater pressure.

RECHARGE AREA An area where water enters the soil and contributes to the groundwater store. Upper slopes and areas with shallow soils are common recharge areas. Recharge is maximised where soils overlie fractured rocks, where soils are highly permeable, where vegetation is shallow-rooted or absent, and when rainfall exceeds evapotranspiration.

RIVER SALINITY River salinity is caused by saline discharges from dryland, irrigation and urban salinity, and aquifers into creeks and rivers.

SALINE SEEPAGE An area where groundwater has brought salts to the surface. Usually occurs in the lower reaches of a catchment.

SALINISATION The process whereby soluble salts accumulate within the soil.

SALINITY A term referring to the salt content in soil or water.

SALT SCALD An area where salt crystals accumulate on the soil surface, suppressing plant growth and often leading to surface soil erosion which can expose saline subsoils.

SODICITY Refers to a soil containing levels of sodium that affect its stability. Sodic soils are dispersible and vulnerable to erosion.

SOIL PERMEABILITY The ability of a soil to absorb and move water through itself.

SOIL PORES Very small spaces between soil particles often occupied by air, water or minerals.

SOIL SPONGE The effect of plant roots in drawing water from the soil, which dries it out and maintains watertable heights. Prior to extensive land clearing, native vegetation provided a soil sponge to a depth of 6 metres or more.

STRATIFICATION The formation of horizontal strata or structured layers of differing characteristics.

URBAN SALINITY Salinity that occurs as a result of urban activities.

WATERLOGGING The process whereby a soil becomes saturated with water and most of the soil pores are filled with water rather than air.

WATERTABLE The upper surface of the groundwater below which the soil or rock pores are saturated

Source: NSW Department of Land and Water Conservation, 2000

FURTHER INFORMATION:
WEBSITE ADDRESSES

NATIONAL GOVERNMENT SITES

CSIRO:

www.csiro.au/index.asp

CSIRO scientists are involved in scientific research on and creating solutions for salinity.

DEPARTMENT OF AGRICULTURE, FISHERIES AND FORESTRY:

www.napswq.gov.au

This Commonwealth government website includes information about the National Action Plan for Salinity and Water Quality, which identifies high-priority areas for dryland salinity and key catchments across Australia.

MURRAY-DARLING BASIN COMMISSION:

www.mdbc.gov.au

The Murray–Darling Basin Initiative was created in 1992 to plan and manage the water, land and other environmental resources of the Murray–Darling Basin. This website includes the Murray–Darling Basin Salinity Audit with predictions for all major river valleys in the basin for the next 20, 50 and 100 years. There is also information on water and land salinity and the Basin Salinity Management Strategy for the next 15 years

NATIONAL DRYLAND SALINITY PROGRAM:

www.ndsp.gov.au

The National Dryland Salinity Program is jointly sponsored by state and Commonwealth governments and rural industries. It involves researching, developing and extending practical approaches to dryland salinity. It is also responsible for publication of *SALT* magazine, which looks at solutions and management of salinity Australia wide.

NATIONAL LAND AND WATER RESOURCES AUDIT:
www.nlwra.gov.au/ANRA/atlas_home.cfm
The results of the National Land and Water Resources Audit are
provided by the Australian Natural Resources Atlas.

PRIME MINISTER'S SCIENCE, ENGINEERING AND INNOVATION COUNCIL:
www.dest.gov.au/science/pmseic/documents/salinity.pdf
This is the web address for the Prime Minister's Science,
Engineering and Innovation Council paper on Dryland Salinity and
its impacts on Rural Industries and the Landscape.

STATE AND LOCAL GOVERNMENT SITES

NEW SOUTH WALES DEPARTMENT OF SUSTAINABLE NATURAL RESOURCES:
www.dlwc.nsw.gov.au/care/salinity/index.html
This NSW government website includes the NSW Salinity Strategy.

QUEENSLAND DEPARTMENT OF NATURAL RESOURCES AND MINES:
www.nrm.qld.gov.au/salinity/index/html
This Queensland government website includes information about
the state's salinity audit and implementation of the National Action
Plan.

**SOUTH AUSTRALIAN DEPARTMENT OF WATER, LAND AND BIODIVERSITY
CONSERVATION:**
www.dwr.sa.gov.au/waterpublications/pdfs/salinityrm.pdf
This website includes the South Australian River Murray Salinity
Strategy 2001–2015.

**TASMANIAN DEPARTMENT OF PRIMARY INDUSTRIES, WATER AND
ENVIRONMENT**
www.dpiwe.tas.gov.au
This Tasmanian government website provides information on land
management in the state, including the audit report and maps on
salinity.

VICTORIAN DEPARTMENT OF SUSTAINABILITY AND ENVIRONMENT:
www.nre.vic.gov.au

This Victorian government website includes information on land and water management and salinity in Victoria.

WAGGA WAGGA CITY COUNCIL:
www.wagga.nsw.gov.au

The city of Wagga Wagga has been a pioneer in tackling urban salinity and has produced information about the cause, effects and treatment of it.

WESTERN AUSTRALIAN DEPARTMENT OF AGRICULTURE:
www.agric.wa.gov.au/environment/land/salinity/

This website includes information on salinity in Western Australia.

ORGANISATIONS

AUSTRALIAN CONSERVATION FOUNDATION:
www.acfonline.org.au

Salinity is one of the environmental campaigns of the ACF.

CO-OPERATIVE RESEARCH CENTRE FOR FRESHWATER ECOLOGY:
www.enterprise.canberra.edu.au/www/www-crcfe.nsf

This website includes information about the centre, which is designed to explore ways of improving the conditions of Australia's inland waters.

COOPERATIVE RESEARCH CENTRE FOR WATER QUALITY AND TREATMENT:
www.waterquality.crc.org.au/

The research centre was established in 1995 to provide national strategic research for the Australian water industry. Its mission is to help the Australian water industry produce high-quality water at an affordable price.

ENGINEERS AUSTRALIA:

www.ieaust.org.au/

This institution (originally known as the Institution of Engineers Australia, or IEAust) represents engineers and their work throughout Australia, some of which includes engineering solutions for salinity management.

IRRIGATION ASSOCIATION OF AUSTRALIA:

www.irrigation.org.au

A site developed by the association to provide information about the irrigation industry and its practices.

LANDCARE

www.landcareaustralia.com.au/

Landcare Australia is a community-based land conservation movement that was created in 1988 by Phillip Toyne, director of the Australian Conservation Foundation, and Rick Farley, director of the National Farmers Federation. Landcare groups tackle a range of land degradation issues including salinity.

NATIONAL FARMERS FEDERATION:

www.nff.org.au/

The NFF is made up of state farm organisations, commodity councils, associates and affiliates. It is responsible for national issues, which affect more than one state or commodity including salinity.

SAVE THE MURRAY:

www.savethemurray.com

Save the Murray is a not-for-profit organisation supported by the Commonwealth and South Australian governments. This website includes facts and history on the Murray River plus practical ideas for saving water.

Water. A sleeper issue and one of the most powerful—and controversial—in Australia today. Here at last is a book that tackles conflicting interests head on.

Punchy, pacy and perceptive, *Watershed* takes us through the fascinating history of water in this country. It examines the rise of water moguls unknown to most of us, and explains why water now pitches state premiers against each other and the Commonwealth. It's full of yarns of battles for water at the grass roots level, past and present, and issues clear warnings of the great environmental challenges facing the nation: dams, salination, pollution and conservation, and clean water for our cities. *Watershed* is a wake-up call for every Australian.

ABC
Books

SALINITY
Australia's Silent Flood

The greatest environmental threat to
Australia in the 21st Century
A 4-part series examining the effect of salinity
on rural and urban Australia

Introduced and narrated by David Wenham

ABC
Video Program
Sales

FOR EDUCATIONAL PURPOSES ONLY
EXEMPT FROM CLASSIFICATION

ABC Video Program Sales
GPO Box 9994 Sydney 2001
Please complete the payment details below

1300 650 587
Fax: 02 8333 3975
progsales@your.abc.net.au

VIDEO ORDER FORM

I wish to order the following quantity of **The Silent Flood**

Quantity	THE SILENT FLOOD 4 x 30 minute VHS PAL format video	Standard/School*	Price $
x	**(Compilation tape x 4 eps)**	$275/$198	
x	Ep 1: **The Story**	$ 88/$ 66	
x	Ep 2: **The Land**	$ 88/$ 66	
x	Ep 3: **The Water**	$ 88/$ 66	
x	Ep 4: **The Future**	$ 88/$ 66	
Firm sale no returns accepted.		**TOTAL:** $	

*Standard Price denotes all purchases by tertiary institutions, libraries, private and public sector organisations. Schools/Individual Price denotes all purchases for primary and secondary schools and home viewers only.

All prices quoted are inclusive of Goods and Services Tax (GST).

Free postage and handling within Australia only.

Additional charges may apply for overseas or urgent deliveries. Please allow up to 15 working days for delivery of video orders. Prices are current at time of publication and subject to change without notice.

METHOD OF PAYMENT:

❑ Invoice (Schools and Government organisations only)
Please fax official Purchase Order together with this Order Form 02 9950 3975

❑ Cheque/ Money Order
Enclosed is my cheque/ money order for $

❑ Credit Card
Please charge $ amount to my   Bankcard   VISA   MasterCard   (circle)

❑ Card No: _ _ _ _   _ _ _ _   _ _ _ _   _ _ _ _   Expiry:___ /___

Name on Card:

❑ Signature: